ß-Glucosidases

ACS SYMPOSIUM SERIES **533**

ß-Glucosidases

Biochemistry and Molecular Biology

Asim Esen, EDITOR

Virginia Polytechnic Institute and State University

Developed from a symposium sponsored
by the Division of Agricultural and Food Chemistry
at the 204th National Meeting
of the American Chemical Society,
Washington, D.C.,
August 23–28, 1992

American Chemical Society, Washington, DC 1993

Library of Congress Cataloging-in-Publication Data

β-glucosidases: biochemistry and molecular biology / Asim Esen, editor

    p.   cm.—(ACS symposium series, ISSN 0097–6156; 533)

"Developed from a symposium sponsored by the Division of Agricultural and Food Chemistry at the 204th National Meeting of the American Chemical Society, Washington, D.C., August 23–28, 1992."

Includes bibliographical references and indexes.

ISBN 0–8412–2697–0

1. Glucosidases—Congresses.

I. Esen, Asim, 1938– . II. American Chemical Society. Division of Agricultural and Food Chemistry. III. American Chemical Society. Meeting (204th: 1992: Washington, D.C.) IV. Title: Beta-glucosidases. V. Series.

QP609.G4B23   1994
574.19′254—dc20   93–25191
                                                  CIP

Foreword

THE ACS SYMPOSIUM SERIES was first published in 1974 to provide a mechanism for publishing symposia quickly in book form. The purpose of this series is to publish comprehensive books developed from symposia, which are usually "snapshots in time" of the current research being done on a topic, plus some review material on the topic. For this reason, it is necessary that the papers be published as quickly as possible.

Before a symposium-based book is put under contract, the proposed table of contents is reviewed for appropriateness to the topic and for comprehensiveness of the collection. Some papers are excluded at this point, and others are added to round out the scope of the volume. In addition, a draft of each paper is peer-reviewed prior to final acceptance or rejection. This anonymous review process is supervised by the organizer(s) of the symposium, who become the editor(s) of the book. The authors then revise their papers according to the recommendations of both the reviewers and the editors, prepare camera-ready copy, and submit the final papers to the editors, who check that all necessary revisions have been made.

As a rule, only original research papers and original review papers are included in the volumes. Verbatim reproductions of previously published papers are not accepted.

M. Joan Comstock
Series Editor

Contents

INDEXES

Preface

THE ENZYME β-GLUCOSIDASE (β-D-glucoside glucohydrolase, EC 3.2.1.21) occurs widely in prokaryotes and eukaryotes. It catalyzes the hydrolysis of aryl- and alkyl-β-D-glucosides as well as glucosides with only a carbohydrate moiety (e.g., cellobiose). This book reviews and discusses the state-of-the-art knowledge about β-glucosidases of a wide spectrum of organisms (mammal, plant, insect, fungus, and bacterium). It also sets the tone for and points out the direction of future research and applications relative to this ubiquitous enzyme.

Current research on β-glucosidases has significant scientific, medical, and economic implications. Human acid β-glucosidase (glucocerebrosidase) has potential in the development of therapeutic and diagnostic procedures that will be useful in the treatment of Gaucher's disease, an inherited disorder caused by the deficiency of acid β-glucosidase localized to the lysosome. Cytosolic neutral β-glucosidase is implicated in the metabolism of pyridoxine-5′-β-D-glucoside, as well as the hydrolysis of β-glucosides ingested along with foods of plant and animal sources. Plant β-glucosidases have been implicated in a variety of growth, productivity, defense, and food and feed toxicity-related reactions such as cyanogenesis. In addition, recent data indicate that the plant enzyme may be involved in the metabolism of plant hormones such as auxin, gibberellin, and cytokinin, whose storage forms occur as β-glucosides and are activated upon cleavage by β-glucosidase.

Cellulose, our most abundant renewable resource on this planet, forms the highest proportion of municipal garbage (e.g., newspaper and other paper products). In fungi and bacteria, the enzyme is involved in cellulose and cellobiose catabolism as part of the cellulase complex. Cellulases break down cellulose to cellobiose, and β-glucosidases hydrolyze cellobiose to two glucose molecules. β-Glucosidase is inhibited by its end product, glucose; the substrate cellobiose accumulates and in turn inhibits the cellulase complex in organisms that rely on cellulose as their carbon source. If the rate-limiting step catalyzed by β-glucosidase in cellulose hydrolysis can be overcome, glucose production from cellulose by enzymatic means should become economically feasible. Thus, fungal and bacterial β-glucosidases appear as ideal candidates for engineering a β-glucosidase to be used as part of the cellulase complex in the industrial-scale conversion of cellulose to glucose.

In the past decade, considerable progress has been made on the molecular biology and biochemistry of β-glucosidases, but much more remains to be done and known. It is hoped that the data, ideas, and insights presented in this book by leading researchers in the field will serve as a launching pad for future research, discoveries, and applications.

I would like to take this opportunity to express my gratitude to the contributing authors, the ACS Division of Agricultural and Food Chemistry, the ACS Books Department, and the National Science Foundation Biochemistry Program for making possible the symposium and the publication of the book.

ASIM ESEN
Department of Biology
Virginia Polytechnic and State University
Blacksburg, VA 24061

January 12, 1993

Chapter 1

β-Glucosidases

Overview

Asim Esen

Department of Biology, Virginia Polytechnic Institute and State University,
Blacksburg, VA 24061

In the past decade, considerable progress has been made on the
molecular biology and biochemistry of β-glucosidases. Some of
these data have been analyzed and discussed in the context of
specific organisms (plants, fungi, bacteria, and humans) and with
respect to specific problems (biomass conversion, cyanogenesis,
host-parasite interactions, Gaucher's disease). This book brings
together under the same umbrella the leading β-glucosidase
researchers with widely different but complementary research
interests. In the next 15 chapters, these researchers present,
review, and discuss the biochemistry and molecular biology of
β-glucosidases from bacteria, fungi, plants, insects, and humans.
They also provide insights into how knowledge on
β-glucosidases may be applied to problems ranging from
biomass production and conversion to the diagnosis and
treatment of Gaucher's disease.

β-Glucosidases (EC 3.2.1.21) catalyze the hydrolysis of glycosidic linkages
in aryl and alkyl β-glucosides and cellobiose and occur ubiquitously in
plants, fungi, animals, and bacteria. Because β-glucosides and
β-glucosidases are ubiquitous in the living world, one expects to find
structural and catalytic properties shared by all β-glucosidases. In fact, a
review of the literature indicates that almost all β-glucosidases have
subunit molecular weights of 55 to 65 kD, acidic pH optima (pH 5-6), and
an absolute requirement for a β-glycoside (i.e., glucoside, and to a much
lesser extent fucoside and galactoside) as substrate. β-Glucosidases from
widely different sources show remarkable similarity in substrate specificity
for glycone (glucose) and some nonphysiological aglycones (e.g.,
nitrophenols and umbelliferone), although they may have widely different
physiological glucosidic substrates with different aglycone moieties. In
general, β-glucosidases from different orders and kingdoms appear to
differ in their specificities for the aglycone (an aryl or alkyl group) linked
to the glucosyl group by a β-glycosidic bond. β-Glucosidases of fungi,
bacteria, humans, and dicotylodenous plants have been shown to be

0097–6156/93/0533–0001$06.00/0

glycosylated, while those of monocots (maize and sorghum) are not. In plants, the β-glucosidases of dicots are localized to the cell wall (1, 2) or protein bodies (3; Chapter 11) while the β-glucosidases of monocots are localized to the plastid (4, 5, 6, Esen and Stetler, in press). In mammals (e.g., humans and mice), acid β-glucosidase is localized to the lysosome while its neutral counterpart is a soluble protein (cytosolic) (Chapter 7). One would also expect that an enzyme such as β-glucosidase that is not involved in a mainstream metabolic pathway would differ considerably in its structure and specificity from organism to organism due to divergent or convergent evolution. This would be especially true among members of different kingdoms as well as among members of the higher taxonomic groups in the same kingdom.

Current research on β-glucosidases has significant scientific and economic implications. Plant β-glucosidases have been implicated in a variety of key metabolic events: growth, productivity, and food and feed toxicity-related reactions. In humans, one β-glucosidase, commonly known as glucocerebrosidase (β-D-glucosyl-N-acylsphingosine glucohydrolase) or human acid β-glucosidase, catalyzes the degradation of glucosylceramide in the lysosome. The deficiency of the enzyme leads to an inherited disease, Gaucher's disease, in which the glycosphingolipid glucosylceramide accumulates within the lysosomes of reticuloendothelial cells, producing one of the three forms of this disease (7, Chapter 6). The molecular biology of the gene encoding this human β-glucosidase and the physicochemical properties of the enzyme have been studied and are well-understood. Human acid β-glucosidase has potential in the development of therapeutic and diagnostic procedures, particularly for the treatment of Gaucher's disease. β-Glucosidases of cellulolytic organisms (some fungi and bacteria) have been the subject of much past and ongoing research. These enzymes are expected to be targets for genetic engineering to design and select β-glucosidases for specific applications such as biomass conversion.

Substrate Specificity and Natural Substrates

One of the controversial issues related to β-glucosidases is their substrate specificity. Are β-glucosidases a sort of "jack of all trades" enzymes catalyzing the hydrolysis of β-glycosidic linkages in all mono and diglycosides? Are they specialists catalyzing only the hydrolysis of β-glycosidic linkages between glucose and specific aryl and alkyl aglycones? Another way of asking the question is: is the enzyme specific for the glycone or the aglycone or both? For example, maize β-glucosidase purified to homogeneity show 5 times greater activity towards p-nitrophenyl (pNP) β-D-fucoside than pNP β-D-glucoside (the difference is due to higher V_{max}; the K_ms are the same). Similarly, its relative activity towards pNP β-D-galactoside is about 10% of that towards pNP β-D-glucoside. Is the maize enzyme a β-fucosidase, a β-glucosidase or a β-glycosidase?

The questions of substrate specificity and the importance of using natural substrates during enzyme purification and characterization have been ably addressed by Conn (Chapter 2). These issues are of utmost importance to those who use and work with plant β-glucosidases where the natural substrate may not be a single glucoside but includes a vast array of aglycone groups, most of which are taxon-specific and linked to glucose through β-glycosidic bonds. Conn describes numerous examples of aglycon specificities from his work and the literature where the use of chromogenic (mostly nitrophenyl) or fluorogenic (4-methyl umbelliferyl) artificial substrates may have led or could lead workers to erroneous conclusions as to the presence or absence of β-glucosidase activity as well as to its substrate specificity.

Although the use of natural substrates is desirable during enzyme purification and characterization when the natural substrate is known and readily available in pure form, this is not often the case. Often one has an enzyme activity based on assays of tissue extracts using common artificial substrates and is trying to purify this activity with the ultimate objective of identifying and characterizing its *in vivo* or natural substrates. In other instances, one has an abundant (e.g. dhurrin of sorghum leaves) substrate or one that is easily measured and identified and is trying to characterize the enzymatic activity (β-glucosidase) catalyzing its hydrolysis.

In any case, the ultimate goal is to dissect and understand a system --the structure and function of the enzyme β-glucosidase and its substrates. A readily available artificial substrate may offer convenience or may be the only one to use in some cases until one purifies and characterizes the enzyme and identifies its natural substrates. One can draw three important conclusions from Conn's chapter. 1) Purify your enzyme to homogeneity before attempting studies on its substrate specificity. The same message is conveyed by a dictum attributed to Nobel Laureate Arthur Kornberg, "Don't waste clean thinking on dirty enzymes". (Interestingly, Kornberg, in his book *For the Love of Enzymes*, attributes this statement to Efraim Racker.) 2) Identify and characterize the natural substrates of the β-glucosidase you have purified or are working with. 3) If you cannot detect β-glucosidase activity in tissue extracts of an organism using the common chromogenic and fluorogenic artificial substrates, be careful not to conclude that the activity is absent.

The Mechanism of Catalysis

Substantial progress is being made in various laboratories using site-directed mutagenesis on fungal, bacterial and human β-glucosidase genes (Chapters 3-6), and there is considerable evidence for the involvement of carboxylate ions in catalysis. However, the active center has not been entirely and unequivocally mapped of any one β-glucosidase. Similarly, the relation of the mechanism of catalysis to the essentially absolute specificity for a β-glycosyl and to the varying specificities for an aglycone or glycone group remains to be elucidated.

As in other enzyme-catalyzed reactions, the mechanism of catalysis is one of the central questions that needs to be addressed for β-glucosidases. It is known that the β-glucosidases from a variety of sources, studied for substrate specificity, show broad glycon and aglycon specificities. This is so whether they normally hydrolyze a specific natural substrate or not. Thus, the data and its interpretation regarding the mechanism of β-glucosidase catalysis have to be consistent with the observed broad glycon and aglycon specificity. Equally important are the mechanism of binding to substrates and the sequence in which binding between the glycone and aglycone portions of the substrate and the enzyme catalytic site takes place. The nature of the interactions between substrate and enzyme during binding and catalysis should be also elucidated.

Clarke et al. (Chapter 3) review the amino acid sequence similarity, kinetic, inhibitor, chemical modification and affinity labelling, and mutagenesis data from studies on β-glucosidases mostly of fungal and bacterial origin. These data strongly support the involvement of acidic amino acids (Glu and/or Asp) as the acid catalyst and stabilizing anion/nucleophile during catalysis. In fact, the involvement of the same amino acid residues in catalysis has been shown or proposed for all β-glycosidases including β-glucanases, xylanases, and lysozyme in addition to β-glucosidases. Clark et al. conclude that β-glucosidases use a double displacement mechanism to hydrolyze their substrates. Two acidic residues (Asp/Glu) at the active center participate in the hydrolysis, and there is a covalently linked enzyme-substrate intermediate. Clearly, further research on β-glucosidases from different sources is needed in order to map the active center and understand the details of the formation of the enzyme-substrate complex, transition state, and the cleavage of the glycosidic bond.

Identification of covalently linked enzyme-transition state intermediates has been used in understanding the mechanism of catalysis by a variety of enzymes. This approach can be used in combination with other promising and productive approaches, namely, site-directed mutagenesis targeting the specific amino acids forming the active site. Trimbur et al. (Chapter 4) have used both of these approaches in order to define the residues critical to catalysis as well as the mechanism of catalysis by the β-glucosidase of *Agrobacterium* sp. Using tritiated 2-4-dinitrophenyl 2-deoxy-2-fluoroglucoside as a mechanism-based inhibitor and peptide sequencing, they show that Glu 358 is the nucleophile which attacks the glycosidic bond of the substrate and to which glucose is covalently linked during catalysis. This glutamate is invariant in the sequence TENG in β-glucosidases belonging to BGA family. Site-directed mutagenesis studies targeting Glu 358 and other conserved (catalytically important) residues around Glu 358 again clearly show that it is the nucleophile responsible for stabilization of the transition state. For example, each of the nine different mutations at Glu 358 reduced the activity 10^4-fold or more, with only the conservative substitution Asp giving measurable, albeit 2500-fold lower, activity. Drastic reductions in activity were also observed when Gly 360 and Asp 374 were mutated. These elegant studies of the Vancouver group strongly indicate that the acidic residues are crucial to the catalytic action of *Agrobacterium* sp β-glucosidase: Glu 358 as the nucleophile in the formation of the covalent enzyme-glycosyl intermediate, and Asp 374 as the acid-base catalyst to protonate the leaving group (glucose) and then deprotonate the water. Another conclusion drawn from these studies is that the substrate (aglycon) binding and catalytic sites are different and distinct because enzyme inactivated by substrate analogues can continue catalysis.

Human and Mammalian *β*-Glucosidases and *β*-Glucosides

Among the mammalian *β*-glucosidases, the human acid *β*-glucosidase or glucocerebrosidase has been the most studied and best characterized both at the protein and gene level mostly due to its role in Gaucher's disease.   The natural substrates for the enzyme in humans are sphingosylglucosides and glucosylceramide.   Accumulation of these substances in the absence of the enzyme is responsible for the three clinical Gaucher phenotypes.   The enzyme also hydrolyzes a variety of artificial substrates. Grabowski et al. (Chapter 6) review their own studies as well as those of others on the enzymology and molecular biology of human acid *β*-glucosidase.

A unique aspect of glucocerebrosidase is that it is the only known membrane-bound *β*-glucosidase and one which requires an activator protein, saposin C. Mutations at the saposin gene produce a nonfunctional activator and cause Gaucher-like symptoms because, in mutant individuals, the acid *β*-glucosidase is not activated. Nearly three-fourths of Gaucher patients are homozygous for the same mutation which changes the amino acid glutamine to serine at position 370. A pseudogene shows 96% sequence similarity to the functional gene. Furthermore, human and mouse acid *β*-glucosidases show 84% similarity at the nucleotide sequence level in the coding region and 88% similarity at the amino acid sequence level suggesting considerable selective pressure for sequence conservation. Yet the enzyme shows little or no similarity to its plant and yeast counterparts.

The human acid *β*-glucosidase gene has been of pharmacological and biotechnological interest because of its potential for producing recombinant enzyme and prescribing it as a supplement for Gaucher patients.   Presently, the enzyme is extracted and purified from human placenta and marketed (Genzyme Inc., Boston) for delivery to patients through intravenous injection. Injections have to be repeated at two-week intervals, and the treatment by this approach is extremely costly, over $300,000 per year per patient (7). Bacteria cannot be used as a host for active recombinant enzyme production because co-translational glycosylation, a requirement for activity, does not occur in bacteria. Attempts to correct the genetic defect by gene therapy are underway and have yielded encouraging results in a mouse model system.

Glew et al. (Chapter 7) review the literature on a mammalian cytosolic *β*-glucosidase which appears to be the *β*-glucosidase with perhaps the broadest known substrate specificity.   This enzyme shows no immunological similarity to acid *β*-glucosidase (glucocerebrosidase). The enzyme hydrolyzes *β*-glycosides having the glycone *β*-D-glucose, *β*-D-galactose, *β*-D-xylose, and α-L-arabinose, and the aglycone moieties p-nitrophenol, 4-methylumbelliferyl and a variety of alkyl groups. Glew et al. emphasize the point that the hydrolysis of p-nitrophenyl α-L-arabinose does not indicate that the enzyme lacks specificity for anomeric configuration of the monosaccharide moiety. What all four monosaccharides substrates have in common is their trans equatorial configuration with respect to the oxygen at positions C_1, C_2 and C_3. The nature of the substituents at the C_4 and C_5 positions of the pyranose ring is also thought to be important to substrate specificity.

Fairly exhaustive studies searching for physiological substrates for the mammalian cytosolic β-glucosidase have also been reviewed and documented by Glew et al. None of the potential substrates such as glycosphingolipids, mucopolysaccharides, glycoproteins, and steroid glucosides have been hydrolyzed to any appreciable extent and therefore considered as physiological substrates.

An interesting property of the mammalian cytosolic β-glucosidase is its ability to catalyze transglycosylation reactions. Glew et al. summarize some of their studies that show alcohols serving as acceptor molecules in transglycosylation reactions. Although transglycosylation can be readily demonstrated *in vitro*, and alcohols stimulate enzyme activity many fold at low concentrations, the relevance of transglycosylation *in vitro* to transglycosylation *in vivo* needs to be demonstrated. Likewise, the nature of the acceptor and donor specificities in transglycosylation reactions and the sizes of the oligosaccharide products from transglycosylations remain to be determined.

Glew et al. raise an intriguing question relative to the *in vivo* function of cytosolic β-glucosidase. Is it possible that the enzyme catalyzes the hydrolysis of a variety of plant β-glucosides contained in foods of plant origin and absorbed by the digestive system? The fact that the cytosolic enzyme hydrolyzes such plant glucosides as vicine, prunasin, amygdalin, and cycasin suggest that it is hydrolyzing these β-glucosides and many others *in vivo* in various organs (e.g. liver, kidney, brain, and intestine). The aglycone moiety of some of these glycosides may be toxic and cause diseases in man and other mammals. It is expected that these hypotheses will be rigorously tested in the future and the enigmatic position of cytosolic β-glucosidase will be clarified.

Gregory (Chapter 8) presents and discusses data from his laboratory on the β-glucosidic conjugates of pyridoxine, especially pyridoxine 5-β-glucoside, and complements it with a thorough review of pertinent literature. Pyridoxine β-glucosides are the major vitamin B_6 derivatives in addition to pyridoxine in many foods of plant origin. They are of significance to human and animal nutrition because they may serve as the source of pyridoxine after hydrolysis by a β-glucosidase. Alternatively, these glucosides may inhibit competitively the absorption and utilization of pyridoxine. If pyridoxine β-glucoside is indeed hydrolyzed, there has to be a β-glucosidase to catalyze this reaction. Gregory postulates that mammalian cytosolic neutral β-glucosidase is the enzyme that hydrolyzes pyridoxine β-glucosides. This hypothesis needs to be rigorously tested because it has the potential not only for elucidating the fate of dietary pyridoxine β-glucosides, but also for identifying a natural substrate for cytosolic neutral β-glucosidase.

An equally important aspect related to pyridoxine β-glucosides is how they are utilized by plants and whether or not they are hydrolyzed by a specific β-glucosidase. Gregory and others have shown that almond β-glycosidase can hydrolyze pyridoxine β-glucosides, suggesting that plants may not need a β-glucosidase whose specificity is restricted to pyridoxine β-D-glucosides. Thus there is need for studies aimed at understanding the metabolism of pyridoxine β-D-glucosides both in plants and animals.

Plant β-Glucosidases and Their Functions

Plant β-glucosidases have been known for over 150 years since the description of the action of emulsin (almond β-glycosidase) on amygdalin, the cyanogenic β-D-gentiobioside of almonds, by Liebig and Wohler in 1837 (8). In plants, β-glucosidases have been implicated in a variety of key metabolic events and growth-related responses. They range from defense against some pathogens and herbivores through the release of coumarins, thiocyanates, terpenes, and cyanide to the hydrolysis of conjugated phytohormones (e.g., glucosides of gibberellins, auxins, abscisic acid, and cytokinins). However, most of the postulated functions attributed to plant β-glucosidases have not been rigorously documented and elucidated. Presently, the primary structure of two plant β-glucosidase (cyanogenic β-glucosidase of white clover and cassava) genes and enzymes is known (Chapter 11).

Poulton (Chapter 12) has separated the emulsin of black cherry (*Prunus serotina*) into three different enzymes activities, amygdalin hydrolase (AH) which catalyzes the breakdown of amygdalin to glucose and prunasin; prunasin hydrolase (PH) which catalyzes the hydrolysis of prunasin to glucose and mandelonitrile; and mandelonitrile lyase which catalyzes the dissociation of mandelonitrile to the toxic products HCN and benzaldehyde. These three enzymes not only catalyze distinct reactions sequentially but they also occur in different subcellular compartments. Sequencing their genes and studying their temporal and spatial expression are expected to shed further light on this fascinating system.

In plants, one of the functions of β-glucosidase is thought to be in cyanogenesis, the release of HCN upon the hydrolysis of cyanogenic glucosides. Cyanogenesis has been shown to occur in over 3000 plant species belonging to 110 different families (9). It was shown in well-studied model systems that the enzyme (β-glucosidase) and the substrates (cyanogenic glucosides) are present in different cellular compartments. The compartmentalization has to be disrupted, as would be the case after injury to cells and tissues by herbivores and pathogens, for the enzyme and the substrate to come in contact and lead to the hydrolysis of cyanogenic glucosides and the release of HCN. This would suggest that cyanogenesis is a chemical defense response to organisms feeding on intact plant parts or attacking the plant through a site of injury. One needs both laboratory and field studies to document unequivocally the involvement of β-glucosidases and their cyanogenic glucoside substrates in deterring either general or specific herbivores and pathogens by producing HCN and other toxic products.

Earlier studies (10) indeed provide evidence for the role of cyanogenesis in defense against certain herbivores. Kakes (Chapter 10) has recently performed field studies to investigate the relationship between cyanogenesis and the grazing behavior of slugs and snails in natural populations of the cyanogenic plant Trifolium repens. He shows that out of the four cyanotypes (Lili Acac, Lili acac, lili Acac, and lili acac; Li and Ac denotes, respectively, the presence of the β-glucosidase linamarase and the substrates linamarin and and lotaustralin) in natural populations, the lili Acac was attacked less frequently than the others indicating that the presence of cyanogenic glucosides alone is sufficient to deter slugs and snails. The presence of the plant linamarase β-glucosidase was not thought to be significant as long as the feeding herbivore had a cyanogenic β-glucosidase in its gut capable of hydrolyzing the cyanogenic glucosides in the plant.

It is apparent that further studies aimed at answering the questions related to β-glucosidase and cyanogenic glucosides are needed. Are both cyanogenic glucoside and β-glucosidase required for cyanogenesis and its defensive function? Which products of cyanogenesis play the more important role in deterrence, HCN or the carbonyl compound released? Is defense function a general or specific response to attackers, and how do the enzyme, substrates and attackers determine the mechanism of response? Do cyanogenic β-glucosidases have broad specificity for β-glucosides or narrow specificity for one or only a few structurally similar cyanogenic glucosides? When specificity is narrow, or when a cyanogenic β-glucosidase is missing what enzyme or enzymes hydrolyze noncyanogenic β-glucosides? How are cyanogenic β-glucosides used and catabolized during plant growth and development in the absence of conditions causing cyanogenesis?

In fact some of the above questions have been addressed by several of the chapters in this book. Selmar (Chapter 13) discusses the metabolism of β-glucosides with different aglycon moieties in vivo. He presents and reviews data indicating that many β-glucosides are transported within the plant as diglycosides which can not serve as substrates for β-glucosidases. This is especially important in systems where β-glucosidase is apoplastic (e.g., cell wall-bound) while its natural substrates are in the vacuole or other cytoplasmic compartments. A β-monoglucoside could not be transported out of the cell because it would be hydrolyzed by the wall-bound enzyme. Yet "safe" transport out of the cell would occur after conversion to a diglucoside if that compound is resistant to hydrolysis. Selmar describes the situation in Hevea brasiliensis where the cyanogenic β-glucoside linamarin is stored in large quantities in the endosperm. During germination and early seedling development the total linamarin content rapidly decreases without any apparent release of HCN suggesting that linamarin is converted to noncyanogenic substances. In fact, Selmar (Chapter 13) and his colleagues showed through precursor feeding experiments that the diglucoside linustatin was taken up by Hevea cotyledons and converted to linamarin and noncyanogenic compounds. The same studies showed that the label introduced into the cotyledon as diglucoside appeared in linamarin in leaves, suggesting strongly that diglucosides are transport forms of the monoglucosides. Selmar proposes that in all plants with apoplastic *β*-glucosidase, stored monoglucosides (e.g. linamarin and lotaustralin) are mobilized and converted to the corresponding diglucosides (e.g. linustatin and neolinustatin) for transport to other plant organs where they are converted back to cyanogenic glucosides and other substances. This would suggest that *β*-glucosides are metabolized to other compounds and have other functions in the plant besides cyanogenesis.

Answers to some questions about cyanogenesis and other processes involving *β*-glucosides and *β*-glucosidases are beginning to come from studies of cloned *β*-glucosidase genes (cDNAs). These studies provide information not only about the primary structure of *β*-glucosidase genes and their protein products but also the extent and genetic basis of their polymorphisms. The molecular studies also provide opportunities to employ the approaches of "reverse genetics" to identify the tissues in which *β*-glucosidase genes are expressed as well as subcellular sites of enzyme sequestering and action. The same approaches are indispensable for efforts directed at reducing or eliminating toxic products of cyanogenesis in foods and feeds by genetic manipulation and screening. The chapter by Hughes (Chapter 11) provides a concise account of cloning and characterization of cyanogenic *β*-glucosidase genes from white clover and cassava. Her data show the presence of at least three different genes in white clover and one in cassava. Furthermore, white clover and cassava *β*-glucosidases share extensive sequence similarity (67%) and they have the presumptive active site motif I/VTENG and other conserved sequences shared with fungal and bacterial *β*-glucosidases. Analysis of gene expression indicates that *β*-glucosidase mRNA and protein are expressed in cassava in multicelled laticifiers; this is not in agreement with the localization of the enzyme to the cell wall in an earlier study (11).

In addition to cyanogenesis, plant *β*-glucosidases have been implicated in the hydrolysis of glucosides of phytohormones and thus activation of these hormones. Campos et al. (Chapter 14) provide data supporting the widely held view that maize *β*-glucosidase might be indeed involved in phytohormone activation and metabolism. Maize *β*-glucosidase is an abundant soluble protein in young, growing plant parts although its precise function and physiological substrates have not been unequivocally identified. The enzyme has been localized in the plastid by immunocytochemical and histochemical studies in our laboratory (6, *Esen and Stetler, in press*). Campos et al. (Chapter 14) show that a 60 kD polypeptide, termed p60, binds to 5-azido-[7-^{3}H] indole-3-acetic acid based on photoaffinity labelling studies using young coleoptile extracts. Photoaffinity labelling of *β*-glucosidase was specific because it was competitively inhibited by synthetic and natural auxins. The authors show that p60 cleaves indole-3-acetyl-*β*-D-glucoside and that it is a *β*-glucosidase based on the comparison of the sequence of its peptides (and that of its sequence deduced from cloned cDNA) with those of other β-glucosidases. It is clear that p60 is the same protein as the one purified and characterized by Esen and Cokmus (12) and Esen (6). In fact, the N-terminal sequence of pm60, an isoform of p60, (Table I, Chapter 14) is identical to that of maize β-glucosidase published earlier by Esen (6). Campos et al. conclude that p60 may play an important role during seed germination and in plant growth by releasing free auxin and other phytohormones from their inactive, glycosidic conjugates. This is the first demonstration of the hydrolysis of an auxin-glucoside by a β-glucosidase although a role for β-glucosidases in the activation of growth hormone derivatives was proposed by previous investigators (13).

Another physiological function for maize β-glucosidase is the hydrolysis of the hydroxamic acid glucoside (DIMBOA-glc), suggested by Cuevas et al., (14) and demonstrated by Babcock and Esen (unpublished). Babcock was able to isolate hydroxamic acid glucosides by chromatography of maize extracts on Sephadex LH-20 and show by paper chromatography that they produce glucose and hydroxamic acid after incubation with purified maize β-glucosidase. Furthermore these studies showed that there was direct correlation between the hydroxamic acid glucoside content of maize organs and parts and their β-glucosidase activity. Since the enzyme is localized in the plastid, it is now necessary to determine the subcellular location of these glucosides of hydroxamic acid and indole so that the locations of enzyme and its substrates can be reconciled.

Insect β-Glucosidases

Although there is a considerable information about β-glucosidases and their substrates from bacteria, fungi, plants, and humans, not much is known about insect β-glucosidases and their substrates. In view of the facts that most insects are herbivores and that insects are believed to be the richest group in terms of species diversity among higher eukaryotes, their β-glucosidase system is open to exploration and may provide new and surprising information. Pioneering studies by Nahrstedt and his co-workers (Chapter 9) are lending support to this prediction.

Nahrstedt and Mueller (Chapter 9) describe a β-glucosidase and its substrates from the larvae of the moth *Zygaena trifolii*. The system is very similar to that of cyanogenic plants belonging to the family Fabaceae. The enzyme is a typical linamarase and its physiological substrates are linamarin and lotaustralin, two well-known cyanogenic glucosides of the plant world. Nearly 85% of the enzyme activity and 33% of substrate occur in the hemolymph along with the enzyme hydroxynitrile lyase. So, the hemolymph of the larvae is a potential "cyanide bomb" (term is Poulton's; see Chapter 12) because it contains the enzymes and the substrates involved in cyanogenesis in the same compartment. The question is: why is cyanogenesis absent in the hemolymph? Nahrstedt and Mueller's data provide the answer: the linamarase is inactive in the intact hemolymph because it is inhibited by the metal ions Mg^{++} and Ca^{++} whose concentrations in the hemolymph are 18.3 and 7.5 mM, respectively. Thus *Zygaena trifolii* linamarase appears to be the first β-glucosidase inhibited by these divalent cations and requires chelators for activation. Cyanogenesis occurs only when the larvae are injured as would be the case after being chewed or eaten by a predator. It appears that *Zygaena trifolii* has a replica of its host's cyanogenic β-glucoside:β-glucosidase system in the hemolymph. In fact, Nahrstedt and Mueller point out that the larvae can synthesize the cyanogenic glucosides and can also sequester them from the host plant parts it eats. These data raise profound evolutionary questions as to how the two similar systems came into being. Answers are likely to be found when the two systems are dissected and analyzed at the molecular level. Thus insect β-glucosidases and β-glucosides are fertile areas for future research and the results from the Nahrstedt Laboratory provide the stimulus for such research.

β-Glucosidase Stability and Activity in Denaturants

Studying the stability and activity of β-glucosidases under conditions that denature and inactivate typical proteins can provide considerable information about the structure of these enzymes. Such information will be useful in selecting and engineering β-glucosidases that can perform their catalytic function under extreme conditions that might be required in the hydrolysis of cellulose (e.g. biomass conversion and waste disposal) on the industrial scale. In addition, β-glucosidase stability and activity under denaturing conditions may allow one to make predictions about the compactness and rigidity of the tertiary structure. Esen and Gungor (Chapter 15) report on the stability and activity of maize β-glucosidase in the presence of such protein denaturants as anionic detergents (SDS and deoxycholate), urea, and organic solvents (methanol, dimethylformamide, and ethylene glycol). Their data indicate that maize β-glucosidase shows considerable stability and activity in the presence of anionic detergents. For example, it retains about 50-70% of its activity up to two days in the presence of 0.1 to 3.2% SDS. It is completely inactivated in the presence of 2 to 5 M urea or 10-30% dimethylformamide. Some of these studies were extended to β-glucosidases from almonds and two fungi (*Trichoderma* and *Penicillium*), they were also found to be active in the presence of denaturants. In fact, *Trichoderma* β-glucosidase was more active than other β-glucosidases in the presence of selected denaturants. Moreover, it was possible to assay enzyme activity in SDS-gels by standard zymogram techniques. Such studies clearly showed that the active form of maize β-glucosidase was a dimer. The course and extent of β-glucosidase stability was influenced by the concentration of denaturant, length of exposure to denaturant, and pH of denaturant medium.

The stability and activity data obtained with β-glucosidases suggest that these enzymes have a compact and rigid structure similar to that of thermophilic enzymes. If this is true, β-glucosidases should be resistant to proteases. In fact this prediction has been confirmed in the case of maize β-glucosidase which retains its structural and functional integrity after exposure to trypsin, chymotrypsin, and proteinase K under both nondenaturing and denaturing conditions (unpublished data). In this regard, thorough and systematic studies are needed to establish the structural basis of the resistance of β-glucosidases to denaturants. Such studies need to be extended to other β-glucosidases in order to determine whether resistance to denaturants and proteases is a common feature of these enzymes.

Fungal β-Glucosidases and Industrial Applications

It is not an overstatement to say that the welfare of humanity will hinge upon our ability to harness the sun's energy either directly or indirectly by extracting energy from the renewable resource cellulose. Cellulose is the most abundant biological macromolecule on earth; 10^{10} to 10^{11} tons of cellulose are synthesized annually by land plants. It represents solar energy converted to chemical energy via the most fundamental biochemical reaction on our planet, photosynthesis. Thus, it is a logical choice for conversion by enzymatic processes to glucose, a universal substrate for metabolism in living systems and a general industrial feedstock and potential renewable energy source. Glucose can serve as starting material for developing carbohydrate foods, alcohol-based fuels, and other commercial products. A time-tested and successful biological process for the hydrolysis of cellulose to glucose is present in some fungi and microorganisms. For example, cellulolytic organisms (e.g. some fungi and bacteria) have an extracellular cellulase complex which catabolizes cellulose to glucose. The extracellular cellulase complex consists of at least three enzymes: endo-β-1,4-glucanase, exo-β-1,4-glucanase, and β-glucosidase. This complex hydrolyzes cellulose to simpler carbohydrates (e.g. cellobiose and glucose). Fungal and bacterial β-glucosidases appear as natural candidates for engineering an ideal β-glucosidase to be used as part of the cellulase complex in the conversion of cellulose to glucose in industrial scale. The rate-limiting step in cellulose hydrolysis is the one that is catalyzed by β-glucosidase, which breaks down cellobiose to two glucose molecules. Because β-glucosidase is inhibited by its end product, glucose, the substrate cellobiose accumulates and in turn inhibits the cellulase complex in organisms such as *Trichoderma*. If the rate of cellobiose hydrolysis could be increased, cellulose would become a more practical raw material for glucose production. This process will require a thorough understanding of the biochemistry and molecular biology of the enzymes catalysing the conversion of cellulose to glucose.

The cellulase complex of the filamentous fungus *Trichoderma viride* has long been a subject of study in industrial and academic laboratories because it presents an ideal model system for developing the enzymatic hydrolysis of cellulose. Maximum hydrolysis of cellulose to glucose takes place when the three enzymes of the cellulase complex are present at optimum ratios. As stated above, the reaction converting cellobiose to glucose, catalyzed by β-glucosidase, is the rate-limiting one. One would like to know what the consequences are to a cellulolytic organism when any of the enzymes of the cellulose complex is missing. Fowler (Chapter

5) has addressed this question by deleting the gene encoding β-glucosidase in *Trichoderma reseei*. The data indicate that β-glucosidase null mutants, which have neither β-glucosidase mRNA nor β-glucosidase protein and activity, can grow on cellobiose or cellulose as their sole carbon source. Thus it appears that the extracellular β-glucosidase is not essential for normal growth and development. The only observed effect of the absence of the β-glucosidase gene was the delay of induction of other cellulase enzymes at the level of transcription for 12-36 hours. However, the data do not explain how the organism can grow on cellulose and cellobiose without a β-glucosidase. It is conceivable that *Trichoderma* has other β-glucosidase loci and β-glucosidases or alternative systems to take up and utilize cellobiose.

In parallel experiments, Fowler deleted the cellobiohydrolase II (cbh2) gene of *T. reseei*. The strain with cbh2 deletion exhibited normal growth on lactose, comparable to that of the parental wild type strain. When grown on Avicel cellulase, the cbh2$^-$ showed 18-24 hours lag in growth. In this case there was no lag in the induction of cellulase genes, but cellulase mRNA (cbh1) levels were reduced indicating changes in transcription rate or in RNA stability. It is clear that further research is needed to understand the regulation of the extracellular cellulase complex and its individual components at levels from transcription to posttranslation.

Woodward et al. (Chapter 16) describe a method of immobilization "in maintenance-free, propylene glycol alginate/bond, gelatin spheres" for *Aspergillus niger* β-glucosidase and its uses in the hydrolysis of cellulose (newsprint). The β-glucosidase immobilized in these spheres had a slightly higher pH optimum, lower energy of activation, and higher thermostability than the native enzyme. The immobilized enzyme, tested over a period of 74 days at 10-100 mM cellobiose concentrations, in a prototype bioreactor, converted cellobiose to glucose with a very high efficiency. More significantly, the immobilized enzyme was used repeatedly and successfully as a supplement to a commercial cellulase preparation in the hydrolysis of cellulose (newsprint). Thus it appears that Woodward et al. have developed and successfully tested the prototype of a future bioreactor to be used in enzymatic hydrolysis of cellulose and cellulosic wastes to glucose at the industrial scale.

Literature Cited

1. Kakes, P. *Planta*. **1985**, *166*, 156-160.
2. Frehner, M.; Conn, E. E. *Plant Physiol.* **1987**, *84*, 1296-1300.
3. Swain, E.; Li, C. P.; Poulton, J. E. *Plant Physiol.* **1992**, *100*, 291-300.
4. Thayer, S. S.; Conn, E. E. *Plant Physiol.* **1981**, *67*, 617-622.
5. Nisius, A.; Ruppel, H. G. *Planta*. **1987**, *171*, 443-452.
6. Esen, A. *Plant Physiol.* **1992**, *98*, 174-182.
7. Beutler, E. *Science*. **1992**, *256*, 794-799.
8. Liebig, J.; Wöhler, F. *Annalen* **1837**, *22*, 11-14.
9. Poulton, J. E. *Plant Physiol.* **1990**, *94*, 401-405.

10. Jones, D. A. *In Cyanide Compounds in Biology: CIBA Foundation Symposium No. 140.* Evered, D.; Harnett, S., Eds.; John Wiley & Sons, Chichester, UK, **1988**, pp. 151-170.
11. Mkpong, O. E.; Yan, H.; Chism, G.; Sayre, R. T. *Plant Physiol.* **1990**, *93*, 176-181.
12. Esen, A.; Cokmus, C. *Biochem. Genet.* **1990**, *28*, 319-336.
13. Smith, A. R.; Van Staden, J. *J. Exp. Bot..* **1978**, *29*, 1067-1075.
14. Cuevas, L.; Niemeyer, H. M.; Jonsson, L. M. *Phytochemistry*, **1992**, *31* 2609-2612.

RECEIVED April 22, 1993

Chapter 2

β-Glycosidases in Plants

Substrate Specificity

Eric E. Conn

Department of Biochemistry and Biophysics, University of California, Davis, CA 95616

The high degree of specificity which plant β-glycosidases exhibit for the aglycone moiety of their substrates is not well recognized. This has resulted from the use of artificial, often chromogenic substrates rather than those occurring natively. While the use in the past of partially purified, rather than homogeneous, enzymes also contributed to this false impression, improved methods and greater interest in these proteins now requires careful study of this property. Many β-glycosidases acting on cyanogenic glycosides exhibit pronounced specificity for the native substrates, but some exceptions exist. Enzymes hydrolyzing other plant glycosides and thioglucosides also exhibit significant specificity, and, as a group, plant β-glycosidases are well designed for metabolizing specific plant glycosides.

In an earlier review, Hösel cited from the literature many examples of plant glycosidases that exhibit a pronounced specificity for secondary (natural) plant heterosides *(1)*. He noted that "an essential prerequisite for the evaluation of these glycosidases is their thorough purification," and also stressed that physiological substrates must be used when monitoring the purification procedure. Subsequently, Hösel and Conn *(2)* emphasized that the role of the aglycone or non-carbohydrate moiety in determining the specificities of these glycosidases for their natural substrates is largely unappreciated and suggested that more attention be directed to this matter.

It was noted in the later review that the use of mixtures of glycosidases rather than homogenous enzyme preparations was a major reason for the view that glycosidases lack aglycone specificity. The ready availability of almond emulsin, an unresolved mixture of β-glucosidases of comparatively low aglycone specificity, has been a significant factor in establishing this impression. In addition, the wide-spread and frequently exclusive use of artificial substrates such as benzyl, nitrophenyl and 4-methylumbelliferyl glycosides (i.e. chromogenic or fluorogenic substrates) during purification and characterization of glycosidases also has contributed to this commonly held but inaccurate view. Because the artificial substrates are easy to assay, quantitative measurements of their velocities of hydrolysis and substrate

0097–6156/93/0533–0015$06.00/0
© 1993 American Chemical Society

affinities were frequently substituted by only a qualitative assessment of the activity of the enzymes on their natural substrates, compounds which are often available only in limited quantities.

To emphasize this point, a major review *(3)* of plant glycosidases summarizing the properties of ten different types of plant glycosidases (i.e. α- and β-galactosidases, glucosidases, and mannosidases; arabinosidases, N-acetyglucosaminidases, β-fructo-furanosidases and trehalases) states that the assay for eight of these groups involves the use of chromogenic substrates, especially of the nitrophenyl type. At the same time, the reviewers, to their credit, identify naturally occurring substrates, often biopolymers, found in plants as the physiologically important substrates acted on by the enzyme, but fail to emphasize the need to use such substrates during characterization of the plant enzymes.

In the sections following, additional examples of glycosidases demonstrating much specificity towards naturally occuring substrates will be presented, together with another plea for greater utilization of natural substrates when characterizing plant β-glycosidases.

Enzymatic Hydrolysis of *(R)*-Amygdalin and *(R)*-Prunasin

The enzymatic hydrolysis of the cyanogenic glycoside amygdalin by an enzyme mixture from almonds known as emulsin was first described in 1837 *(4)*. Later in this century, several studies were designed to determine how many enzymes are involved in catabolism of amygdalin *(5-10)*. While Armstrong *(5)* and Haisman and Knight *(6)* proposed that two enzymes catalyze the stepwise removal of the two glucose residues from amygdalin, the properties of these two glycosidases, known as amygdalin hydrolase (AH) and prunasin hydrolase (PH), could not be defined since they were not homogeneous. Even today, this matter remains unresolved for the almond enzymes.

Fortunately, there is now accurate information on the mechanism of hydrolysis of amygdalin by a group of three enzymes found in the seeds of the black cherry *(Prunus serotina* L.). In another chapter in this volume, Poulton *(11)* has reviewed their careful work on these three proteins. Their data on the specificity of the two glycosidases involved is a good example of the substrate specificity of plant β-glycosidases *(12)*.

Two homogeneous forms of amygdalin hydrolyase (AH I and AH II) that occur in the cherry seed hydrolyze the terminal glucose of amygdalin and form the monoglucoside known as prunasin *(13)*. These enzymes also hydrolyze *p*- and *o*-nitrophenyl-β-D-glucosides (PNPGlc and OPNGlc) at somewhat reduced rates, and *p*-nitrophenyl-β-D-galactoside (PNPGal) more slowly. OPNGlc, PNPGlc and PNPGal were hydrolyzed by AH I at 84%, 68% and 16% respectively of the rate of amygdalin hydrolysis. While 4-methylumbelliferyl-β-D-glucoside (MUGlu) was also rapidly hydrolyzed (relative rate 110 compared to amygdalin, 100) *(R)*-prunasin and its epimer *(S)*-sambunigrin were not acted upon. AH I does slowly catalyze the hydrolysis of the two cyanogenic glycosides found in flax seed, linustatin and neolinustatin; their relative rates of hydrolysis were 10 and 27% of that of amygdalin. Both isoforms were inactive toward the following glycosides: *(S)*-dhurrin, linamarin, laminarin, cellobiose, sucrose, lactose, maltose, methyl-β-glucoside and phenyl-β-D-glucoside. Gentiobiose was also not hydrolyzed by AH I or II, which is interesting, since amygdalin is a β-gentiobioside.

Three isoforms of prunasin hydrolase, PH I, PH IIa, and PH IIb, have been purified to homogeneity from the black cherry and their substrate specificities examined *(14)*. They show a high degree of specificity for *(R)*-

prunasin and are completely inactive toward the cyanogenic diglucosides *(R)*-amygdalin, linustatin and neolinustatin, as well as the cyanogenic monoglucosides *(S)*-dhurrin and linamarin. Again these enzymes do hydrolyse PNPGlu, OPNGlu and PNPGal at comparable or somewhat reduced rates, as well as MUGlu, but they are inactive toward β-gentiobiose, laminarin, cellobiose, sucrose, lactose, maltose, methyl-β-D-glucoside and phenyl-β-D-glucoside.

Enzymatic Hydrolysis of Linustatin and Linamarin by Flax Glycosidases

A similar study of the enzymatic hydrolysis of a pair of related di- and monoglucosides (linustatin and linamarin) that occur in flax seed *(Linum usitatissimum L.)* again demonstrated the substrate specificity of the related β-glycosidases *(15)*. Linustatinase, which catalyzes the removal of the terminal glucose unit of linustatin forming glucose and the related monoglucoside linamarin, was purified to homogeneity from flax seed and characterized. Linamarase was also purified to homogeneity from the same source. Linustatinase of course hydrolyzed linustatin (relative velocity, 100) and also hydrolyzed neolinustatin, a second cyanogenic diglucoside found in flax seed, with a relative rate of 59. This diglucosidase also hydrolyzed amygdalin, β-gentiobiose and β-1,4-cellobiose with relative rates of 66, 13 and 14 but, as expected only slowly hydrolyzed linamarin and was inactive on *(R)*-prunasin. Linustatinase surprisingly did hydrolyze the monoglucoside dhurrin (relative rate, 167) possibly because the aromatic aglycone bears some slight similarity to the *p*-nitrophenyl group; PNPGlc was also rapidly hydrolyzed by linustatinase (relative rate, 242), perhaps for the same reason.

Kinetic data on the linamarase purified from flax seed is discussed below, together with linamarases from other sources, but for comparison with its related diglucosidase, it was essentially inactive on linustatin (relative rate, 1.3), gentiobiose (relative rate, 2.3) and did not hydrolyze amygdalin. It did rapidly hydrolyze *(R)*-prunasin (relative rate 140) when compared with linamarin (relative rate, 100). It also rapidly hydrolyzed PNPGlc (relative rate, 184) and PNPGal (relative rate, 83) but was only slightly active towards dhurrin (relative rate, 10).

This study demonstrated that linamarase from flax seed possesses somewhat more flexible requirements for the aglycone moieties of naturally occurring substrates as well as chromogenic compounds. In this regard it differs significantly from the dhurrinase of *Sorghum bicolor* Moench *(16)* and the triglochinase of *Triglochin maritima* L. *(17)* that clearly have more stringent requirements.

Substrate Specificity of Linamarase from Several Plants

Enzymes catalyzing the hydrolysis of linamarin forming acetone cyanohydrin and HCN have been purified to apparent homogeneity from several plant sources in recent years *(18-23)*. Some of these are qualified to be called linamarase; the others can not be so named using criteria which are preferred by this author. The interest in this enzyme results, in part, from the fact that all of the plants studied are species of significant economic interest and one, cassava (*Manihot esculenta* Crantz), is a major source of dietary carbohydrate for several hundred million people in Africa and elsewhere.

Table I compares the relative maximal hydrolytic rates (V_{max}) toward several substrates by homogeneous enzyme preparations from cassava petiole *(18)*, butter bean, also known as lima bean *(Phaselous lunatus L.)* seed *(19)*, leaves of *Hevea brasiliensis* (A. Juss.) Muell. Arg. *(20)* and the flax seed linamarase discussed in the previous section *(15)*. The enzymes from flax seed *(15)* and cassava petiole *(18)* qualify to be called linamarase because the native substrate linamarin was used to monitor the purification, including fractions from column chromatographic procedures. The enzyme from butter beans also qualifies as a linamarase, although the substrate employed during purification work was PNPGlc *(19)*. As shown in Table II however, the authors of this study demonstrated that the ratio of the two hydrolytic activities, PNP-β-glucosidase:linamarin β-glucosidase (i.e. linamarase) remained constant throughout the purification of the enzyme from homogenate to the final fraction which was purified 11,700-fold. While it is possible to be misled by these ratios, the most reasonable interpretation is that Itoh-Nashida *et al.* *(19)* did indeed purify linamarase with their procedure.

This criterion of a constant value of the ratio of PNPGlc-ase to linamarase activity was also applied in a recent purification of linamarase from *Trifolium repens* L. *(21)* based on the earlier work of Hughes and Dunn *(22)* which yielded a homogeneous enzyme. Information on this enzyme preparation is not included in Table I, however, because the authors have not examined the ability of the purified enzyme to hydrolyze any naturally occurring substrate other than linamarin. The V_{max} for the *T. repens* enzyme is 2.8 x 10^{-2} kat/kg for linamarin and 5.8 x 10^{-2} kat/kg for PNP-Glc *(21)*. Therefore the chromogenic substrate is hydrolyzed twice as rapidly as the natural substrate by the linamarase for *T. repens*.

The enzyme purified to homogeneity from leaves of *H. brasiliensis* shown in Table I requires additional comment *(20)*. In their study of this enzyme, Selmar et al. provide evidence for it being the only β−glycosidase in the leaves of *Hevea* and therefore prefer to call it a "non-specific β−glycosidase". It could well be called linamarase because linamarin exhibited the highest V_{max} of the 10 naturally occurring glycosides and 10 artificial glycosides examined. Lotaustralin, a homologue of linamarin that occurs in *H. brasiliensis* (and presumably in the other species listed in Table I) was also rapidly hydrolyzed with a V_{max} half that of linamarin; this rare substrate is known to be hydrolyzed by other linamarase preparations. O-nitrophenyl-β−D-glucoside exhibited approximately the same V_{max} as linamarin and PNPGlc was also efficiently utilized as a substrate

It should be pointed out, however, that the only other natural substrates examined, other than linamarin and lotaustralin, whose hydrolysis by this "non-specific glycosidase" might be physiologically significant in *Hevea*, were the diglucosides linustatin and neolinustatin and coniferin, a precursor of lignin in most plants. (Prunasin, amygdalin and dhurrin do not occur in *Hevea*). The V_{max} values of linustatin, neolinustatin and conferin relative to that of linamarin (100) were 10^{-4}, 1.1 x 10^{-4} and 0.043, respectively; i.e. 4 to 6 orders of magnitude less than the V_{max} of linamarin. It is not surprising that the enzyme does not hydrolyze the disaccharides linustatin and neolinustatin. While there may be a need to postulate that this "non-specific" glycosidase, as the *only* glycosidase present in the leaves, must be responsible for hydrolyzing coniferin during lignin synthesis in leaves of *Hevea*, there would seem to be the possibility that a "coniferase" might have been overlooked in this study. Certainly for it to carry out that role, there needs to be a lot of the enzyme present. One wonders, therefore, whether the *Hevea* enzyme is accurately described as a "non-specific β−glycosidase".

Although we have stated that the flax enzyme exhibits less stringent requirements for the aglycone moiety of its substrates, that description is based primarily on its ability to hydrolyze prunasin (and the two chromogenic glycosides PNPGlc and PNPGal) at comparable rates. In this property the flax seed linamarase differs significantly from the cyanogenic β−glycosidases from *Sorghum* and *Triglochin maritima* discussed earlier. Note however that the cassava petiole linamarase (Table I) does not hydrolyze prunasin and that the *Hevea* enzyme also shows little activity toward prunasin. All of the enzymes in Table I do hydrolyze PNPGlc effectively and in this property they clearly differ from the *Sorghum* dhurrinase, but the physiological significance of this property, if there is one, is unknown.

Two other β−glucosidase preparations have been purified from cassava leaves. One of these *(23)* was homogeneous after approximately 100-fold purification and was shown to efficiently hydrolyze linamarin (relative rate, 100 at 10 mM concentration) as well as *(R)*-prunasin (relative rate, 65). It was inactive on amygdalin but like the flax enzyme did hydrolyze PNPGlc (relative rate, 65) and ONPGlc (relative rate, 19). While the enzyme isolated in this study has been well characterized, it does not qualify for inclusion in Table I since the natural substrate linamarin was not used during the purification procedure. The other preparation from cassava leaves *(24)* did not involve measurements of the enzyme's activity toward linamarin at any time and will not be discussed further. To the authors' credit, they did not describe their protein as a linamarase.

Substrate Specificity of Enzymes that Hydrolyze Glucosinolates.

The glucosinolates (mustard oil glucosides) are non-volatile thioglucosides distributed widely in the family Crucifereae; at least 75 have been chemically characterized *(25)*. Plants containing these compounds also contain glucohydrolases that catalyze the hydrolysis of the thioglucoside bond. The aglycone dissociates forming chemically reactive thiocyanates, isothiocyanates, oxazolidine-2-thiones and organic nitriles that contribute the characteristic odors associated with this plant family. These end products also have deleterious effects when they occur in animal feeds *(26)*.

The thioglucosidases (myrosinases) which catalyze the hydrolysis of glucosinolates have been studied for many years, but only a few publications report the catalytic properties of homogeneous preparations of the enzyme. Björkman and Lönnerdal *(27)* reported the properties of one isoenzyme from white mustard seed *(Sinapis alba L.)* which was obtained in a homogeneous state. Their data can be compared with that recently obtained in Poulton's laboratory on a homogeneous preparation of myrosinase from cress seedlings *(Lepidium sativum L.) (28,29)*.

Durham and Poulton have carried out the most extensive study of substrate specificity of myrosinase *(29)* yet reported, although they utilized only two glucosinolates, sinigrin and benzyl glucosinolate, in their work because of the rarity of compounds of this sort from commercial or other sources (Table III). The K_m of the cress enzyme for sinigrin, benzylglucosinolate and PNPGlc were 300 μM, 295 μM and 2.0 mM respectively. Other authors have reported K_m values in the range of 30-360 mM for sinigrin, and 2 mM for PNPGlc; see reference *28* for details. The corresponding V_{max} values for the cress enzyme for were 123, 204 and 86 μmol/hr/mg enzyme for sinigrin, benzylglucosinolate and PNPGlc respectively. These values were obtained in the absence of ascorbic acid which influences both the V_{max} and K_m values of myrosinases of most species for their substrates.

Table I. Relative Maximal Hydrolytic Activities (V_{max}) with Different Substrates
for Linamarase from Several Sources [a]

	Linum usitatissimum[15] seed[b]	Manihot esculenta[18] petiole	Phaseolus lunatus[19] seed	Hevea brasiliensis[20] leaves
Linamarin	100	100	100	100
Linustatin	1.3	--	--	10^{-4}
Prunasin	140	0	51	0.25
Amygdalin	0	0	0.6[c]	3×10^{-5}
Dhurrin	10	--	--	5×10^{-3}
p-Nitrophenyl-β-D-Glc	184	441	112	36
p-Nitrophenyl-β-D-Gal	83	0	75	2

[a]Number at the species name indicates reference and source of data. [b]Relative activities
measured at 6.25 mM. [c]Activity relative to linamarin, both measured at 10 mM.

Table II. Purification of β-Glucosidase from Butter Beans (200 g)[a]

Procedures	Total units (U)	Specific activity (U/mg)	Ratio[b]
Homogenate	740	0.0155	1.99
(NH4)2SO4 fractionation	526	0.0247	1.92
Acetone fractionation	497	0.107	2.06
CM-Sephadex C-50	477	1.690	2.04
DEAE-Sephadex A-50	462	119.1	2.11
Sephadex G-200	373	181.1	2.09

[a]Adapted with permission from Table I in ref. 29. [b]Ratio of the activities, PNP-β-
glucosidase/linamarin β-glucosidase.

Table III. Substrate specificity of cress myrosinase

Substrate	Relative activity[a]
Sinigrin	100
Benzylglucosinolate	102
p-Nitrophenol-β-D-glucoside	18
o-Nitrophenol-β-D-glucoside	2

[a]Expressed as a percentrage of sinigrase activity (100% = 70
μmol/h/mg protein). Reprinted with permission from ref. 29.
Copyright 1990.

The relative rates of hydrolysis of sinigrin, benzylglucosinolate, and two chromogenic substrates PNPGlc and PNPGal at 1 mM are shown in Table III. The data show that the two glucosinolates are hydrolyzed at equal rates while the chromogenic glucosides are much less rapidly hydrolyzed. What is important about these data is that these four glucosides are the only ones which were hydrolyzed by this purified enzyme out of a group of 29 examined. Those which were not hydrolyzed by the cress enzyme were: (R)-amygdalin, (R)-prunasin, salicin, methyl-β−D-glucoside, phenyl-β−D-glucoside, gentiobiose, cellobiose, sucrose, lactose, maltose, p-aminophenyl-1-thio-β−D-glucoside, PNP-thio-β−D-glucoside, PNP-β−D-galactoside, xyloside and mannoside; N-acetylglucoaminide and N-acetylgalactosaminide; PNP-α−D-glucoside, galactoside and mannoside; ONP-β−D-galactoside and xyloside; 4-methylumbelliferyl-β−D-glucoside and mannoside; and 4-methylumbelliferyl-α−D-glucoside. Clearly the cress seedling is an enzyme that exhibits significant substrate specificity.

The substrate specificity for compounds other than glucosinolates were not reported for the white mustard myrosinase studied by Björkman and Lönnerdal (27). They did show that their one homogeneous isoenzyme hydrolyzed the following six glucosinolates: glucotropaeolin, sinigrin, glucocheirolin, progoitrin, glucosinalbin and glucocapparin. It is not possible to calculate the relative velocities for hydrolysis of these compounds since the authors used only 2 concentrations in the range of 0.05 to 0.3 mM and the rates were not linear. The K_m of the homogeneous isozyme for sinigrin was 0.17 mM. These authors did not examine the ability of their enzyme to hydrolyze chromogenic substrates.

Hydrolysis of Chromogenic and Fluorogenic Substrates

In this paper the ability of plant glycosidases to hydrolyze chromogenic and fluorogenic substrates has been deemphasized. The main reason for doing so is that studies which claim to examine the action of plant glycosidases on naturally occurring glycosides should utilize those compounds during the purification of the enzyme and must examine the natural product as a substrate. There are now several examples of plants containing glycosidases which utilize artificial substrates but which are inactive on the endogenous glycoside of interest. Similarly, there are examples of glycosidases which utilize the naturally occurring substrate of interest, but have little or no activity on chromogenic substrates. A good example is the mixture of glycosidases found in a buffered extract of etiolated and light grown seedlings of *Sorghum bicolor* (16).

Figure 1 shows the separation of a protein extract of 5-day-old etiolated sorghum seedlings on a DE-52 column (gradient starting at fraction 60). The graph shows that a protein fraction containing an enzyme that hydrolyzes PNPGlc (line connected by closed triangles) but inactive toward dhurrin passes through the column and appears in fractions 10-30. After the gradient starts, a double peak containing two proteins which hydrolyze dhurrin is eluted (line connected by closed circles) but these fractions do not hydrolyze PNPGlc. Finally, a third peak appears which hydrolyzes both dhurrin and PNPGlc, apparently in a constant ratio. Clearly this graph shows that if one were to use PNPGlc to follow dhurrinase activity, the investigator would miss about 80% of the dhurrinase activity in these seedlings. In addition, the presence of the PNPGlc-ase that can not hydrolyze dhurrin would surely complicate matters during the purification.

In light-grown seedlings (Figure 2) the fractions before the start of the gradient appear to contain two PNPGlc-ases, neither of which act on dhurrin. This tissue, however, lacks the dhurrinase which is inactive on PNPGlc and does appear to have two peaks of an enzyme acting on both dhurrin and PNPGlc.

Figure 3 shows a similar situation in which an extract of leaves of *Trifolium repens* was placed on a chromotofocussing column and eluted *(30)*. The graph shows that there are three fractions eluted at pH 7, pH 6 and pH 4.7 (line connected by closed squares) which hydrolyze 4-methylumbelliferyl-b-D-glucoside but have little or no activity toward linamarin (line connected by closed circles). A final fraction eluted at pH 4.3 does hydrolyze both linamarin and the fluorogenic glucoside.

The three examples shown in Figures 1-3 are from studies on plants containing cyanogenic glycosides, To demonstrate that this situation is not unique for cyanogenic glycosidases and their natural substrates, consider the data in Figure 4 taken from reference *31*. This figure shows the separation on a DEAE-cellulose column of an extract of stratified apple seeds which was examined for glucosidase activity against the phenolic glycoside phloridzin (solid line) and the chromogenic glucoside PNPGlc (dotted line). The graph shows that three "peaks" (5a, 5b and 4) contain enzymes that possess PNPGlc-ase activity and phloridzinase activity in a ratio of approximately 100-120 to 1. The remaining "peak" on the other hand exhibits a ratio of approximately 1.0. Here again the extensive use of PNPGlc as a substrate for monitoring phloridzinase activity during purification would clearly be unwise. In addition, the fact that phloridzin is commercially available at reasonable cost also should persuade enzymologists interested in this type of problem to forego the use of artificial glycosides.

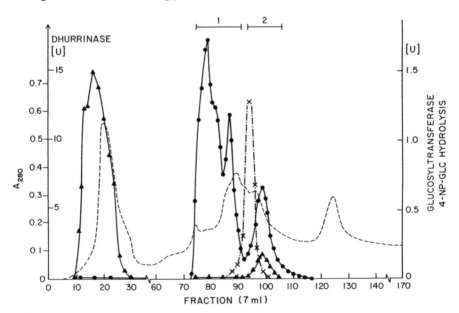

Figure 1. Separation of a protein extract of shoots of 5-day-old dark-grown sorghum seedlings on a DE-52 column. The gradient starts at fraction 60. ---, protein (A_{280}); ●, dhurrin hydrolysis; ▲, 4-NPG hydrolysis; -·-·-, 4-HMN:UDPG glucosyltransferase; |—1—|, dhurrinase 1 (combined and further purifed; |—2—|, dhurrinase 2 (combined and further purified). (Reproduced with permission from ref. 16. Copyright 1987 Academic Press.)

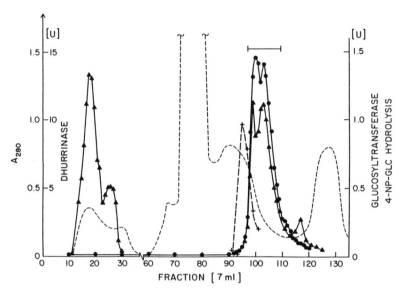

Figure 2. Separation of a protein extract of 7-day-old light-grown sorghum shoots on a DE-52 column. The conditions and symbols are those described for Fig. 1; I---I , fractions combined and further purified. (Reproduced with permission from ref. 16. Copyright 1987 Academic Press.)

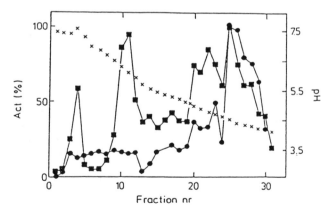

Figure 3. PBE 94 chromatofocussing of an extract of plant Tr 4500-24. Both linamarase (●—●) and 4 methylumbelliferyl-β-D-glucosidase (■—■) activity were determined. x—x, pH. (Reproduced with permission from ref. 30. Copyright 1983 Koninklyke Neerla Botanisch Vereniging.)

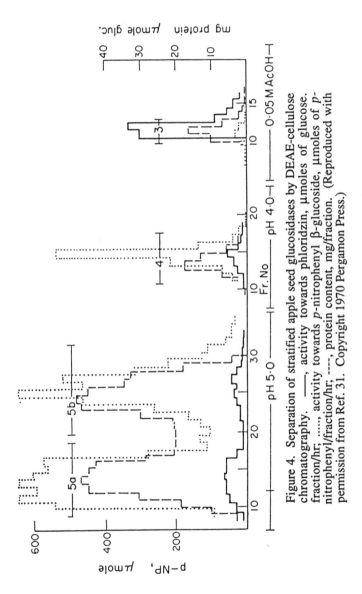

Figure 4. Separation of stratified apple seed glucosidases by DEAE-cellulose chromatography. ——, activity towards phloridzin, μmoles of glucose. fraction/hr; ……, activity towards p-nitrophenyl β-glucoside, μmoles of p-nitrophenyl/fraction/hr; ----, protein content, mg/fraction. (Reproduced with permission from Ref. 31. Copyright 1970 Pergamon Press.)

Summary

The aim of this paper is to emphasize that plant β-glycosidases exhibit high specificity for the aglycones of their substrates. This is essential if these catalysts are to perform the metabolism of their naturally-occurring, glycosylated substrates. Possessing such specificity, the plant β-glycosidases do not differ from other enzymes of intermediary metabolism which are well "designed" to perform a specific role in the metabolism of the organism in which they occur.

Literature Cited

1. Hösel, W.; In: *Biochem. Plants*, Conn, E.E., Ed. Academic, New York, NY, **1981**, Vol. 7, pp. 725-753.
2. Hösel, W; Conn, E.E. *Trends Biochem. Sci.* **1982**, 7, 219.
3. Dey, P.M.; Del Campillo, E.I. *Adv. Enzymol.* **1984**, 56, 141.
4. Liebig, J.; Wohler, F. *Annalen.* **1837**, 22, 11.
5. Armstrong, H.E.; Armstrong, E.F.; Horton, E. *Proc. Roy Soc. B.* **1912**, 85, 359.
6. Haisman, D.R.; Knight, D.J. *Biochem. J.* **1966**, 103, 528.
7. Haisman, D.R.; Knight, D.J.; Ellis, M.J. *Phytochem.* **1967**, 6, 1501.
8. Lalegerie, P. *Biochemie* **1974**, 56, 1163.
9. Helferich, F.; Kleinschmidt, T. *Hoppe-Seyler's Z. Physiol. Chem.* **1966**, 340, 31.
10. Grover, A.K.; MacMurchie, D.D.; Cushley, R.C. *Biochim. Biophys. Acta* **1977**, 482, 98.
11. Poulton, J.E. In: (this volume) Esen, A., ed.; ACS Sym. Series No. ___ **1993**, ___.
12. Kuroki, G.W.; Lizotte, P.A.; Poulton, J.E. *Z. Naturforsch.* **1984**, 39C, 232.
13. Kuroki, G.W.; Poulton, J.E. *Arch. Biochem. Biophys.* **1986**, 247, 433.
14. Kuroki, G.W.; Poulton, J.E. *Arch. Biochem. Biophys.* **1987**, 255, 15.
15. Fan, T.W.-M.; Conn, E.E. *Arch. Biochem. Biophys.* **1985**, 243, 361.
16. Hösel, W.; Tober, I.; Eklund, S.H.; Conn, E.E. *Arch Biochem. Biophys.* **1987**, 252, 152.
17. Nahrstedt, A.; Hösel, W; Walther, A. *Phytochem,* **1979**, 18, 1137.
18. Eksittikul, T.; Chulavatnatol, M. *Arch. Biochem. Biophys.* **1988**, 266, 263.
19. Itoh-Nashida, T.; Hiraiwa, M.; Uda, Y. *J. Biochem.* **1987**, 101, 847.
20. Selmar, D.; Lieberei, R.; Biehl, B; Voigt, J.; *Plant Physiol.* **1987**, 83, 557.
21. Pocsi, I.; Kiss, L.; Hughes, M.A.; Nanasi, P., *Arch. Biochem. Biophys.* **1989**, 272 496.
22. Hughes, M.A.; Dunn, M.A.; *Plant Mol. Biol.* **1982**, 1, 169.
23. Mkpong, O.E.; Yan, H.; Chism, G.; Sayre, R.T. *Plant Physiol.* **1990**, 93, 176.
24. Yeoh, H.-H; Sia, H.-L. *Physiol. Plant.* **1989**, 11, 187.
25. Larsen, P.O. In: *Biochem. Plants* Conn, E.E., Ed.; Academic, New York, NY **1981**, Vol. 7, pp. 501-525.
26. Tookey, H.L.; Van Etten, C.H.; Daxenbichler, M.E., In: *Toxic Constituents of Plants*, 2nd ed; Liener, I.E., Ed., Academic, New York, NY **1980**, pp. 103-142.

27.   Björkman, R.; Lönnerdal, B. *Biochem. Biophys. Acta* **1973**, *327*, 121.
28.   Durham, P.L.; Poulton, J.E. *Plant Physiol.* **1989**, *90*, 48.
29   Durham, P.L.; Poulton, J.E. *Z. Naturforsch.* **1990**, *45C*, 173.
30.   Boersma, P.; Kakes, P.; Schram, A.W. *Acta Bot. Neerl.* **1983**, *320*, 39.
31.   Podstolski, A.; Lewak, S. *Phytochem.* **1970**, *9*, 296.

RECEIVED January 12, 1993

Chapter 3

β-Glucosidases, β-Glucanases, and Xylanases
Their Mechanism of Catalysis

A. J. Clarke, M. R. Bray, and H. Strating

Department of Microbiology, University of Guelph, Guelph, Ontario
N1G 2W1, Canada

The pathway for the enzymatic hydrolysis of carbohydrates involves
either one chemical transition state or two, depending upon
whether the configuration at the anomeric center of the sugar
product is inverted or retained, respectively. The cellulolytic β-
glucosidases, together with many β-glucanases and xylanases, are
retaining enzymes implying that they operate by the double
displacement mechanism. Here, we review the evidence obtained
by amino acid sequence homology studies, kinetic considerations,
chemical modification and affinity labeling experiments, and site-
directed mutagenesis for the participation of Glu/Asp residues
acting in concert as the acid catalyst and stabilizing anion/
nucleophile in the mechanism of action of the cellulolytic enzymes.

The β-(1→4) glycoside hydrolases, like all enzymes, are grouped and classified
according to their specificities and action patterns. As a class of enzymes
however, the β-glucosidases (EC 3.2.1.21) display very broad specificity with
respect to both the aglycon and glycon moieties of substrates. For example,
natural substrates include the steroid β-glucosides and β-glucosylceramides of
mammals, cyanogenic β-glucosides of plant secondary metabolism in addition to
the cellulose of plant cell walls. These enzymes have attracted the attention of
biochemists and bioorganic chemists alike in view of the simplicity relative to
other β-glycosidases of both their mode of action and their substrates. These
initiatives have greatly contributed to our understanding of the β-glycosidases and
yet their mechanism of action remains a major topic of debate. Much of the
mechanistic detail has been delineated from the study of the cellulolytic β-
glucosidases of bacteria and fungi.
    The microbial degradation of cellulosic biomass requires the action of at
least three groups of enzymes; cellulase (EC 3.2.1.4), cellobiohydrolase (EC
3.2.1.91) and β-glucosidase which act synergistically to hydrolyse the β-1,4 bonds

0097–6156/93/0533–0027$06.00/0
© 1993 American Chemical Society

of cellulose to glucose (*1*). β-Glucosidases bind soluble cello-oligosaccharides in one productive mode and hydrolyse glucosyl residues sequentially from the non-reducing end. The substrate-binding region of the *Trichoderma reesei* enzyme is composed of three to four subsites with the catalytic groups located between subsites one and two (*2*) and kinetic studies with the *Aspergillus niger,*(*3*) *Pyricularia oryzae* (*4*) and *Sclerotium rolfsii* (*5*) enzymes would suggest this is a common feature of all cellulolytic β-glucosidases. As described below, the mode of catalysis amongst the β-glucosidases, and possibly the majority of the β-glycosidases, is also the same.

Stereochemistry

The most important mechanistic aspect of β-glucosidases, and indeed of all glycosidases, is the stereochemistry of the catalysed reaction. Hydrolysis may occur with either retention or inversion at the anomeric center of the reducing sugar product, and this will dictate whether one chemical transition state or two are involved in the pathway. β-Glucosidases, together with most cellulases, cellobiohydrolases and xylanases, are known to hydrolyse their substrate with retention of the anomeric configuration (*6*). This was first shown by Parrish and Reese (*7*) with the β-glucosidase from sweet almond emulsin by a rapid derivatization of enzyme-substrate reaction mixtures and the determination of the anomeric configuration of the glucose adduct by gas-liquid chromatography. A direct approach employing high-field NMR to monitor the chemical shift and coupling constant of the anomeric proton on the hemiacetal carbon of the product glucose has since confirmed this result (*8*). Mechanistically, this implies that these enzymes operate by the double displacement mechanism of Figure 1. The basic tenets of this mechanism were first proposed by Koshland 40 years ago (*9*), and as recently reviewed by Sinnott (*10,11*) its revised form involves the following features:

i)   an acid catalyst which protonates the substrate;
ii)   a carboxylate group of the enzyme positioned on the opposite side of the sugar ring to the aglycone;
iii)   a covalent glycosyl-enzyme intermediate with this carboxylate in which the anomeric configuration of the sugar is opposite to that of the substrate;
iv)   this covalent intermediate is reached from both directions through transition states involving oxocarbonium ions;
v)   various non-covalent interactions providing most of the rate enhancement.

Amino Acid Sequence Homologies

The complete and partial amino acid sequences of over 300 glycosyl hydrolases are now known, of which approximately 85 are β-1,4-glucanases and xylanases. Comparison and alignment of these latter sequences together with hydrophobic cluster analysis revealed considerable homology between fungal and bacterial

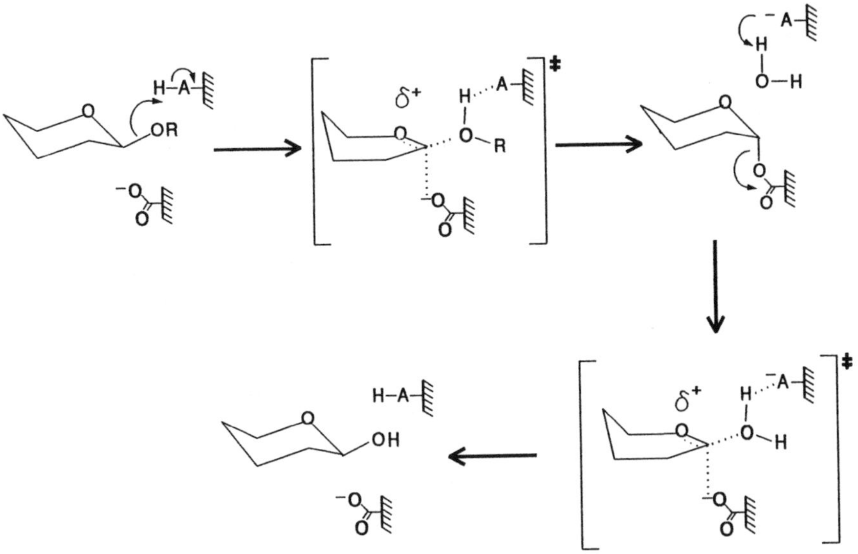

Figure 1. Generally accepted endocyclic pathway of the double displacement mechanism proposed for the retaining β-glycosidases.

enzymes and enabled the establishment of classification schemes comprising 6, (12), 7 (13) or 8 (14) families. An analogous system was subsequently developed to include all the glycosyl hydrolases (15). Of the 35 distinct families, 10 involve the β-glucanases and xylanases with the β-glucosidases comprising Families 1 and 3. Thus, with the paucity of detailed structural and mechanistic data, it is not surprising that predictions of catalytic residues and domains for the β-gluco-sidases were made based on homology studies with other well studied β-glycosidases such as hen egg white lysozyme (HEWL).

Evidence for carboxylic acids serving as the acid catalyst and stabilizing anion/nucleophile in glycosidases is founded primarily on the X-ray crystallo-graphic structure of HEWL (16,17). Glu-35 of HEWL, with pK values of 6.1 and 6.7 for the free enzyme and enzyme-substrate complex, respectively, was identified as the proton donor during catalytic hydrolysis while Asp-52 serves to stabilize the putative oxonium transition state. The unusually high pK values for Glu-35 have been attributed to; i) the low accessibility of solvent to the active-site cleft, ii) electrostatic interactions with the charge constellation (mainly Asp-52), and iii) its location at the negative site (C-terminus) of an α-helix macrodipole (18). Site-directed mutagenesis experiments have since supported the ascribed role of Glu-35 in HEWL (19,20).

Interesting amino acid sequence homologies have been found between the region of HEWL comprising Glu-35 and Asp-52 and segments of various β-glycosidases (21-25), including β-glucosidases (23,25) (Figure 2). By analogy with HEWL, Moranelli et al. (25) identified Glu-160 and Asp-177 of the Schizophyllum commune β-glucosidase and Glu-626 (or Asp-629) and Asp-644 of the Candida pelliculosa enzyme as potential catalytic residues. With the Agrobacterium sp. enzyme, the putative active site sequence was proposed to involve Glu-311 and Asp-327 (23). These proposals were based on the strong conservation of the residues since very little other homology was found between these different enzymes; indeed the β-glucosidases involved belong to different families of the various classification schemes. However, with the delineation of more amino acid sequences, the proposed lysozyme-like region was not found to be generally conserved within respective families of enzymes, which severely discounts the significance of the observed homologies.

A highly conserved region has been observed among the aligned sequences of 16 different cellulases which contains an -Asn-Glu-Pro- motif (Figure 3). Site-directed mutagenesis of two of the genes encoding cellulases that display very little other homology supported the prediction that the Glu residue was essential for activity (26). This motif was subsequently shown to exist in the Family 1 β-glucosidases (27) and the homologous region was extended to include conserved Arg and His residues (Figure 3). The roles of these latter residues were investigated by site-directed mutagenesis of the Clostridium cellulolyticum cellulase A gene (28). The Arg residue is speculated to either participate in a salt bridge that maintains the structural organization of the enzyme or hydrogen bond to the substrate transition state. The His-122 residue is thought to be more essential for catalysis, but it is unlikely the proton donor since substituting His-122 with Ser, Glu, Gly or Phe did not totally abolish catalytic activity. The His residue identified in the catalytic center of the C. thermocellum cellulase by

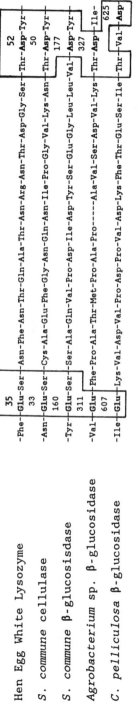

Figure 2. Amino acid sequence alignment of a fungal cellulase and both fungal and bacterial β-glucosidases with the segment of hen egg white lysozyme containing its catalytic residues Glu-35 and Asp-52.

Agrobacterium sp.	77	GVEAY**R**FSLAWPRIIPDGFGPINEKGLDFYDRLVDGCKARGIKTYATL**YH**W**D**LPLTIMGD
C. saccharolyticum	69	GLKAY**R**FSIAWTRIFPDGFGTVNQKGLEFYDRLINKLVENGIEPVVTL**YH**W**D**LPQKLQDI
B. polymyxa	72	GIRTY**R**FSVSWPRIFPNGDGEVNQEGLDYYHRVVDLLNDNGIEPVVTL**YH**W**D**LPQALQDA
C. thermocellum	72	GIKSY<u>**R**FSISWPRIFPE</u>G<u>TGKLNQKGLDFYDRLTNLLENGIMPAITL**YH**W**D**LPQKLQ</u>DK

Agrobacterium sp.	137	GGWASRSTAHAFQRYAKTVMARLGDRLDAVATFN**E**P
C. saccharolyticum	129	GGWANPEIVNYYFDYAMLVINRYKDKVKKWITFN**E**P
B. polymyxa	132	GGWGNRRTIQAFVQFAETMFREFHGKIQHWLTFN**E**P
C. thermocellum	132	GG<u>WKNRD</u>IT<u>DYFTEYSE</u>VIFKNLGDIVPIW<u>FTHNE</u>P

Figure 3. Sequence alignments of Family A β-glucosidases (BGA of Béguin (13)) and identification of conserved essential residues. Underlined residues denote regions of significant homology (at least 3 of 4) and the essential residues, Arg, His and Glu, are both in bold face and enlarged.

chemical modification experiments (His-516) was likewise concluded not to be the proton donor following site-directed mutagenesis studies (*29*). While it is possible that the Glu of the *Asn-Glu-Pro* motif serves as the proton donor during catalysis, the affinity labeling and trapping experiments described below have identified other Glu/Asp residues as the nucleophilic partner.

Kinetics and Inhibitor Studies

The presence of both an acid catalyst and a stabilizing anion/nucleophile in *β*-glucosidases has been inferred from pH-activity dependence studies. The hydrolysis of a wide variety of both natural and synthetic substrates catalysed by the enzymes from *Aspergillus wentii* (*30*), *Botryodiplodia theobromae* (*31,32*), *S. commune* (*33*), *S. rolfsii* (*5*), *Thermoascus aurantiacus* (*34*), *Trichoderma viride* (*35*), and sweet almonds (*36*) show a bell-shaped dependence on pH. The simplest explanation for this pH dependence is that two catalytically-essential groups on the enzyme can ionize, but activity occurs only when one is protonated and the other is not. The apparent pK values for the essential ionizable groups in these enzymes calculated from these plots were 3.2 - 4.4 for the acidic limbs and 5.3 - 6.9 for the basic limbs (Table I). While it was obvious that the ionizable group(s) with the more acidic pK values were carboxylates, the identification of the more basic residues was not as facile. Early workers naturally invoked the participation of a His residue (*31,32,34*), but chemical modification experiments with group specific reagents (see below) have precluded their direct role in catalysis (*33,37*).

Table I. Comparison of apparent pK values for essential ionizable group of free enzyme (e) and enzyme-substrate complexes (es)

Enzyme Source	pK_{e1}	pK_{es1}	pK_{e2}	pK_{es2}	*Reference*
β-Glucosidases					
A. wentii A$_3$	-	-	-	6.1	30
B. theobromae	3.50	3.65	6.0	6.75	31,32
T. aurantiacus	-	3.19	-	5.64	34
Sweet Almonds	4.4	-	6.7	-	36
T. viride	-	3.5	-	6.8	35
S. commune	3.3	3.3	6.6	6.9	33
Cellulase					
S. commune	3.7	3.8	6.1	6.6	38
Lysozyme					
Hen egg white[a]	3.7	3.7	6.1	6.7	17

[a]Values of pK_1 and pK_2 are for Asp-52 and Glu-35, respectively.

Detailed kinetic studies have further supported the influence of two carboxylic acids on the catalytic activity of β-glucosidase. Dale *et al.* (*36*) determined the second-order rate constants (k_{cat}/K_m) for the β-glucosidase-catalysed hydrolysis of a series of aryl-β-D-glucopyranosides. The ΔH_{ion} values near zero for the two ionizable groups on the sweet almond enzyme (pKs 4.4 and 6.7) suggested the nature of the groups as carboxylic acid/carboxylate. In addition, earlier results obtained with reversible inhibitors by the same researchers had hinted that the protonated group with a pK of 6.7 is a neutral acid (*39*).

Despite the evidence given above, the importance of acid catalysis on aglycone departure has been questioned. No solvent isotope effect on k_a ($k_a(H_2O)/k_a(D_2O)$) was observed for the β-glucosidase-catalyzed hydrolysis of a number of aryl glucosides suggesting little proton transfer to the leaving group (*36*). Indeed, an analogous residue to the putative acid catalyst of HEWL, Glu-35, is not readily evident in the crystal structure of T4 phage lysozyme (*40*). However, a physical-organic study of the *A. wentii* A$_3$ enzyme indicated that the catalytic proton was almost completely transferred at the rate-determining transition state of similar substrates (*41*).

Whereas there is conflicting evidence for the importance of a carboxylic acid serving as the acid catalyst in β-glucosidases, the converse is true for the existence of an essential carboxylate at the active center. The pH-activity studies strongly suggest the presence of an anion at the active center of β-glucosidase and this is further supported by investigations involving various competitive inhibitors of the enzyme.

Since the early observations of Larner and Gillespie (*42*) that Tris is an inhibitor of several glycosidases, including β-glucosidase, kinetic studies have shown that glucosylamines are high affinity competitive inhibitors of the enzyme. D-Glucosylamine binds to the β-glucosidase 4 orders of magnitude more tightly than does D-glucose (*43*). Using a series of *N*-substituted β-glucosylamines and glycosylpyridinium salts, both the sweet almond (*44*) and *A. wentii* (*41*) β-glucosidases were shown to be more strongly inhibited by basic glucosyl derivatives than by the corresponding neutral analogues. By determining the ratio of the inhibition constants of the respective compounds, the additional interaction energy ($\Delta \Delta G°$) provided by the basic or cationic species was calculated to range from 13 to 23 kJ•mol^{-1}. A more recent investigation with reversible inhibitors (phenols and amines) of the sweet almond enzyme confirmed and extended the earlier studies (*39*). These data led each group to conclude that glucosyl cation-like transition states were involved in substrate hydrolysis and an anionic group (a carboxylate ion) is present at the active site of the enzymes. This Glu/Asp residue is implicated in stabilizing the positive charge of the putative oxocarbonium ion transition state and in addition likely forms a covalent intermediate (Figure 1). Indeed, the observed α-deuterium kinetic isotope effect for hydrolysis by the *Stachybotrys atra* β-glucosidase is consistent with the formation of a covalent glucosyl-enzyme intermediate (*45*).

Chemical Modifications

Chemical modification experiments employing a variety of reagents have since confirmed the essential role of only carboxylic acids/carboxylates in the catalytic mechanism of the *S. commune* β-glucosidase while discounting any role for His residues (*33*). In this study, the most effective chemical modification reagent tested was found to be a specific carbodiimide which was used at pH 6.0. Since it is known that carbodiimides react only with the protonated form of carboxylic acids (*46*), the concomitant loss of activity with modification can be ascribed to the presence of an essential Asp/Glu residue with a high pK value (6.6 - 6.9 for the *S. commune* β-glucosidase). Protection from inactivation was provided by the competitive inhibitor deoxynojirimycin which blocked the modification of three residues. Two of these are likely the catalytically essential residues while the third could be involved with substrate binding. Reaction of the enzyme with diethylpyrocarbonate resulted in only a partial loss (22%) of catalytic activity indicating that it is highly unlikely that the essential ionizable group exhibiting pK values near neutrality may be ascribed to a His residue. The direct participation of a His residue on the *A. wentii* enzyme was likewise discounted. Photooxidation in the presence of Rose Bengal and reaction with diazotized 5-amino-1*H*-tetrazole failed to completely inactive the enzyme (*37*).

The essential role of carboxylic acids at the active centers of the other cellulolytic and hemicellulolytic enzymes has also been demonstrated by carbodiimide modifications. Protection studies with appropriate ligands for the *S. commune* cellulase (*38*) and xylanase (*47*) indicated that the enzyme-substrate complexes cannot be inactivated by the carbodiimide. That diethylpyrocarbonate served to modify all of the His residues present in these enzymes without significant impairment of activity precludes their involvement in the catalytic mechanism of action of these enzymes.

Further evidence for a catalytic carboxylic acid in the *S. commune* β-glucosidase and cellulase is given by the apparent irreversible binding of transition metals to these enzymes with its concomitant inactivation (*33,48*). The inactivation was speculated to proceed by chelation involving carboxyl groups at the active center in a manner analogous to that observed by both X-ray crystallography and NMR for metal-HEWL complexes (*49,50*).

Affinity Labelling

Confirmation of a carboxylate anion/nucleophile at the active site of the β-glucosidases has been obtained with the mechanism-based affinity label conduritol-B-epoxide and its brominated derivative developed by Legler (*51,52*). These compounds have been used to help delineate the mechanism of action of the enzymes from *A. wentii*, (*53-56*), *A. oryzae* (*56,57*), *B. theobromae* (*58*), bitter (*59*) and sweet almonds (*60*), yeast (*56*), snail (*61*), and human Gaucher cells (*62,63*). In those cases tested, the stereochemistry of inactivation was demonstrated with the release of the epoxide label by hydroxylamine as (+)-*chiro*-inositol which would be expected for a retaining glycosidase (*54,55,60*). Kinetic considerations (*55,60*) and protection studies with substrates and inhibitors (58,62) indicate that conduritol-B-epoxide specifically binds to the catalytic sites of the enzymes and using tritiated derivatives of the compound, inactivation of the various β-glucosidases was found to proceed with the binding of 1 equivalent of affinity label (55,59,60,62,63). Finally, the bound nucleophile has been determined for the A. wentii (53,54), bitter almond (59) and human (62) enzymes to be an Asp residue (Figure 4).

Sigmoidal pH-dependencies of inactivation rates have been observed for the A. wentii (55) and sweet almond (60) β-glucosidases treated with conduritol-B-epoxide. Values of pKs between 5.6 and 7.3 were determined indicating that, possibly unlike typical substrates, the inactivation mechanism does indeed require an acid catalyst to protonate the relatively inert epoxide prior to nucleophilic attack. In contrast, however, the inactivation of the B. theobromae Pat enzyme appears to be dependent upon an acidic group with a pK value of 4.3 (58).

Trapping experiments have been conducted to address the question of whether a covalent intermediate is formed during catalysis or the carboxylate anion simply stabilizes the oxocarbonium intermediate as originally proposed for HEWL. Slow substrates (p-nitrophenyl-β-D-2-deoxyglucoside and D-glucal) were used with the A. wentii β-glucosidase to generate a steady-state concentration of the glycosyl-enzyme intermediates which were subsequently stabilized by guanidinium chloride denaturation. One molecule of the trapped substrate was found to remain bound to the respective enzyme preparations and protease digestion followed by purification of the labeled peptide identified the binding sites as the same Asp residue that binds conduritol-B-epoxide (64,65).

All of these affinity labeling and trapping studies have suggested the correct orientation and participation of an aspartyl residue in stabilizing a substrate intermediate but as pointed out recently (66), they suffer from two major flaws; i) none of these intermediates are catalytically competent, and ii) trapping experiments do not distinguish between strong ion-pair intermediates and covalent linkages. Elegant work by Withers and co-workers has circumvented these issues with the development of a new class of mechanism-based inhibitors. They developed a series of 2-deoxy-2-fluoroglucosides with good leaving groups (eg. dinitrophenolate or fluoride) to trap a covalent intermediate in the normal catalytic mechanism of the β-glucosidases (66-69). It is thought that the inductive effect of fluorine at C-2 slows both glucosyl-enzyme formation and hydrolysis while the reactive leaving group accelerates only the glucosylation which results in the accumulation of the glucosyl-enzyme adduct. These 2-deoxy-2-fluoroglucosides have served as affinity labels for the Alcaligenes faecalis (67-69), Agrobacterium sp. (66) and sweet almond β-glucosidases (68), and with each, protection from rapid inactivation was provided by competitive inhibitors. The enzyme intermediates are stable enough to allow for ^{19}F NMR analysis which showed that the sugar is covalently linked to the A. faecalis (69) and Agrobacterium sp. (66) enzymes via an α-anomeric linkage, as expected for a retaining glycosidase. With the Agrobacterium sp. enzyme, Glu-358, a conserved residue among the Family BGA enzymes (13) (Figure 5), was identified as the bound nucleophile (66). Furthermore, and perhaps more importantly, the catalytic competence of this intermediate was proven with the measurements of rates of reactivation of inactivated enzyme upon subsequent

A. wentii A₃   V-M-S-**D**-W-A-A-H-H-A-G-V-S-G-A-L

Bitter Almonds   I-T-Z-Z-G-V-F-G-**D**-S-(A,B₂,Z,P)-K

Human Gaucher   G-S-Q-R-V-G-L-A-S-Q-K-N-**D**-L-D-A-V-A-L-M-H-P-D-G-S-A-V-V-V-L-N-
                         R-S-S-K-D-V-P-P-T-I-K-A-P-A-V-G-F-L-E

Figure 4. Amino acid sequences of peptides labeled with [³H]conduritol-B-epoxide isolated from inactivated β-glucosidases. With each, the inhibitor was shown to be covalently linked to the Asp residue shown in bold face (53,54,59,62). The same residue of the *A. wentii* enzyme was shown to bind D-glucal (64) and *p*-nitrophenyl-β-D-2-deoxyglucoside (65) after denaturation/trapping experiments.

Agrobacterium sp.   320   **WEVYAPALHTLVETLYERYDLPECYITENGACYNMGV-ENGEVNDQPRLDY**

C. saccharolyticum   312   **WEVFPQGLFDLLIWIKESYPQIPIYITENGAAYNDIVTEDGKVHDSKRIEY**

B. polymyxa   315   **WPVESRGLYEVLH-YLQKYGNIDIYITENGACINDIVTEDGKVQDDRRISY**

C. thermocellum   315   **WIIYPEGLYDLLMLLDRDYGKPNIVISENGAAFKDEIGSNGKIEDTKRIQY**

Figure 5. Sequence alignment of Family A β-glucosidases (BGA (13)) around the conserved Glu residue of the *Agrobacterium* sp. residue shown to covalently bind the slow substrate dinitrophenyl-β-2-deoxy-2-fluoroglucoside (66). Underlined residues reveal residues of significant homology (at least 3 of 4 residues).

addition of an acceptor molecule, glucosylbenzene (*66*). Hence, these studies have provided the first direct evidence for the formation of a covalent intermediate to a carboxylate on a catalytically competent β-glucosidase.

The Mechanism of Catalysis

The experimental evidence described above supports the generally accepted double displacement mechanism (Figure 1) of Koshland (*9*) for the retaining β-glucosidases. Only Asp and Glu residues have been shown to serve as the acid catalyst and stabilizing anion/nucleophile. Thus, the carboxylic acid is thought to protonate the exocyclic oxygen of the substrate to make it a good leaving group, which is stabilized by the correctly positioned carboxylate.   To accommodate unfavourable electronic geometry, the enzymes are postulated to contribute distortion of the protonated substrate. Displacement of the exocyclic group follows with the formation of a glucosyl-enzyme adduct involving a carboxylate nucleophile, which is in turn displaced with H_2O.

While there is little doubt for the formation of an enzyme-bound intermediate, the exocyclic pathway described above for its formation, although widely accepted, is strictly conjecture. Recent kinetic studies on the inhibition by hydroxylated cyclic amines (*70*) and theoretical molecular dynamic calculations of lysozyme (*71*) led to the proposal of an endocyclic pathway for the formation of the enzyme-substrate adduct. Thus, this alternative pathway postulates the protonation of the ring (endocyclic) oxygen by the acid catalyst followed by the nucleophilic attack and displacement (Figure 6). Rather than invoking substrate distortion for the favourable alignment of electronic orbitals, this mechanism requires the rotation of the C_1-C_2 bond of the carbohydrate substrate. These controversial studies have initiated considerable debate but interestingly, as pointed out by Frank in a recent review (*72*), much of the published data on the mechanism of β-glycoside hydrolysis are equally consistent with both pathways. As the intermediate observed by Withers *et al.* (*66,69*) using ^{19}F NMR for the β-glucosidase-catalyzed hydrolysis of glucosyl fluorides can only be reasonably derived from the exocyclic cleavage pathway, he further proposes that these

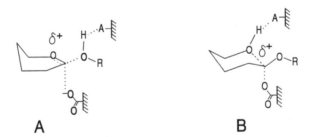

A   B

Figure 6. Proposed transition states of β-glucosidase catalysed hydrolysis by the (A) endocyclic, and (B) exocyclic pathways for the double displacement mechanism leading to the formation of the covalent enzyme-substrate intermediate.

pathways may not be mutually exclusive and the β-glycosidases may catalyze both endo- and exocyclic pathways depending on the substrate.

In conclusion, the β-glucosidases appear to hydrolyze their substrates with the retention of configuration through the concerted participation of two Asp/Glu residues. Subtle details of the nature of the transitions state(s) leading to the formation of the covalent enzyme-substrate intermediate have, however, yet to be elucidated.

Acknowledgments

These studies were supported by an operating grant from the Natural Engineering Research Council of Canada.

Literature Cited

1. Eriksson, K.-E. L.; Blanchett, R.A.; Ander, P. *Microbial and Enzymatic Degradation of Wood and Wood Components*, Springer Series in Wood Science; Springer-Verlag: Berlin, Germany, 1990; pp. 89-180.
2. Chirico, W.J.; Brown, Jr., R.D. *Eur. J. Biochem.* **1987**, *165*, 343-351.
3. Cole, F.E.; King, F.W. *Biochim. Biophys. Acta* **1964**, *81*, 122-129.
4. Hirayama, T.; Horie, S.; Nagayama, H.; Matsuda, K. *J. Biochem. (Tokyo)* **1978**, *84*, 27-37.
5. Shewale, J.G.; Sadana, J. *Arch. Biochem. Biophys.* **1981**, *207*, 185-196.
6. Gebler, J.; Gilkes, N.R.; Claeyssens, M.; Wilson, D.B.; Béguin, P.; Wakarchuk, W.W.; Kilburn, D.G.; Miller, Jr., R.C.; Warren, R.A.J.; Withers, S.G. *J. Biol. Chem.* **1992**, *267*, 12559-12561.
7. Parrish, F.W.; Reese, E.T. *Carbohydr. Res.* **1967**, *3*, 242-429.
8. Withers, S.G.; Dombroski, D.; Berven, L.A.; Kilburn, D.G.; Miller Jr., R.C.; Warren, R.A.J.; Gilkes, N.R. *Biochem. Biophys. Res. Commun.* **1988**, *139*, 487-494.
9. Koshland, D.E. *Biol. Rev.* **1953**, *28*, 416-436.
10. Sinnott, M.L. In *Enzyme Mechanisms*; Royal Society of Chemistry, London, 1987; pp. 259-297.
11. Sinnott, M.L. *Chemical Reviews* **1990**, *90*, 1171-1201.
12. Henrissat, B.; Claeyssens, M.; Tomme, P.; Lemesle, L.; Mornon, J.-P. *Gene* **1989**, *81*, 83-95.
13. Béguin, P. *Annu. Rev. Microbiol.* **1990**, *44*, 219-248.
14. Gilkes, N.R.; Henrissat, B.; Kilburn, D.G.; Miller, R.C. Jr.; Warren, R.A.J. *Microbiol. Rev.* **1991**, *55*, 303-315.
15. Henrissat, B. *Biochem. J.* **1991**, *280*, 309-316.
16. Blake, C.C.F.; Mair, G.A.; North, A.T.C.; Phillips, D.C.; Sarma, V.R. *Proc. Roy. Soc. London, Ser. B.* **1967**, *167*, 365-377.
17. Imoto, T.; Johnson, L.N.; North, A.T.C.; Phillips, D.C.; Rupley, J.A. In *The Enzymes*, 3rd edn; Boyer, P.D., Ed.; 1972, Vol 7; pp 666-868.
18. Spassov, V.; Karshikov, A.D.; Atanasov, B.P. *Biochim. Biophys. Acta* **1989**, *999*, 1-6.

19. Malcolm, B.A.; Rosenberg, S.; Corey, M.J.; Allen, J.S.; deBaetselier, A.; Kirsch, J. *Proc. Natl. Acad. Sci. USA* **1989**, *86*, 133-137.
20. Anand, N.N.; Stephen, E.R.; Narang, S. *Biochem. Biophys. Res. Commun.* **1988**, *153*, 862-868.
21. Paice, M.G.; Desrochers, M.; Rho, D.; Jurasek, L.; Roy, C.; Rollin, C.F.; DeMiguel, E.; Yaguchi, M. *Biotechnology* **1984**, *2*, 535-539.
22. Yaguchi, M.; Roy, C.; Rollin, C.F.; Paice, M.G.; Jurasek, L. *Biochem. Biophys. Res. Commun.* **1983**, *116*, 408-411.
23. Wakarchuk, W.W.; Greenberg, N.M.; Kilburn, D.G.; Miller, Jr., R.C.; Warren, R.A.J. *J. Bacteriol.* **1988**, *170*, 301-307.
24. Morosoli, R.; Roy, C.; Yaguchi, M. *Biochim. Biophys. Acta* **1986**, *870*, 473-478.
25. Moranelli, F.; Barbier, J.R.; Dove, M.J.; MacKay, R.M.; Seligy, V.L. *Biochem. Int.* **1986**, *12*, 905-912.
26. Baird, S.D.; Hefford, M.A.; Johnson, D.A.; Sung, W.L.; Yaguchi, M.; Seligy, V.L. *Biochem. Biophys. Res. Commun.* **1990**, *169*, 1035-1039.
27. Gräbnitz, F.; Seiss, M.; Rucknagel, K.P.; Staudenbauer, W.L. *Eur. J. Biochem.* **1991**, *200*, 301-309.
28. Belaich, A.; Fierobe, H.-P.; Baty, D.; Busetta, B.; Bagnara-Tardif, C.; Gaudin, C.; Belaich, J.-P. *J. Bacteriol.* **1992**, *174*, 4677-4682.
29. Tomme, P.; Chauvaux, S.; Béguin, P.; Dhurjati, P.; Aubert, J.-P. *J. Biol. Chem.* **1991**, *266*, 10313-10318.
30. Legler, G. *Hoppe-Seyler's Z. Physiol. Chem.* **1967**, *348*, 1359-1366.
31. Umezurike, G.M. *Biochem. J.* **1979**, *179*, 503-507.
32. Umezurike, G.M. *Biochem. J.* **1977**, *167*, 831-833.
33. Clarke, A.J. *Biochim. Biophys. Acta* **1990**, *1040*, 145-152.
34. Bedino, S.; Testore, G.; Obert, F. *Ital. J. Biochem.* **1986**, *35*, 207-220.
35. Maguire, R.J. *Can. J. Biochem.* **1977**, *55*, 19-26.
36. Dale, M.P.; Kopfler, W.P.; Chait, I.; Byers, L.D. *Biochemistry* **1986**, *25*, 2522-2529.
37. Legler, G.; Gilles, H. *Hoppe-Seyler's Z. Physiol. Chem.* **1970**, *351*, 741-748.
38. Clarke, A.J.; Yaguchi, M. *Eur. J. Biochem.* **1985**, *149*, 233-238.
39. Dale, M.P.; Ensley, H.E.; Kern, K.; Sastry, K.A.; Byers, L.D. *Biochemistry* **1985**, *924*, 3530-3539.
40. Weaver, L.H.; Matthews, B.W. *J. Mol. Biol.* **1987**, *193*, 189-199.
41. Legler, G.; Sinnott, M.L.; Withers, S. *J. Chem. Soc. (London), Perkin II* **1980**, *1980*, 1376-1380.
42. Larner, J.; Gillespie, R.E. *J. Biol. Chem.* **1956**, *233*, 709-726.
43. Walker, D.E.; Axelrod, B. *Arch. Biochem. Biophys.* **1978**, *187*, 102-107.
44. Legler, G. *Biochim. Biophys. Acta* **1978**, *525*, 94-101.
45. van Doorslaer, E.; van Opstal, O.; Kersters-Hilderson, H.; DeBruyne, C.K. *Bioorg. Chem.* **1984**, *12*, 158-169.
46. Chan, V.W.F.; Jorgensen, A.M.; Borders, C.L. *Biochem. Biophys. Res. Commun.* **1988**, *151*, 709-716.
47. Bray, M.R.; Clarke, A.J. *Biochem. J.* **1990**, *270*, 91-96.
48. Clarke, A.J.; Adams. L.S. *Biochim. Biophys. Acta* **1987**, *916*, 213-219.
49. Perkins, S.J.; Johnson, L.N.; Machin, P.A.; Phillips, D.C. *Biochem. J.* **1979**, *181*, 21-36.

50. Dobson, C.M.; Williams, R.J.P. In *Metal-Ligand Interactions in Organic Chemistry and Biochemistry*; Pullman, B. and Goldblum, N., Eds.; Part 1; D. Reidel Publishing Co., Dordrecht, The Netherlands, 1977; pp. 255-282.
51. Lalégerie, P.; Legler, G.; Yon, M. *Biochemie*, **1982**, *64*, 977-1000.
52. Legler, G. *Mol. Cell. Biochem.* **1973**, *2*, 31-38.
53. Bause, E.; Legler, G. *Biochim. Biophys. Acta* **1980**, *626*, 459-465.
54. Bause, E.; Legler, G. *Hoppe-Seyler's Z. Physiol. Chem.* **1974**, *355*, 438-442.
55. Legler, G. *Hoppe-Seyler's Z. Physiol. Chem.* **1968**, *349*, 767-774.
56. Legler, G. *Hoppe-Seyler's Z. Physiol. Chem.* **1966**, *345*, 197-214.
57. Legler, G. *Hoppe-Seyler's Z. Physiol. Chem.* **1968**, *349*, 1488-1492.
58. Umezurike, G.M. *Biochem. J.* **1987**, *241*, 455-462.
59. Legler, G.; Harder, A. *Biochim. Biophys. Acta* **1978**, *524*, 102-108.
60. Legler, G. *Hoppe-Seyler's Z. Physiol. Chem.* **1970**, *351*, 25-31.
61. Donsimoni, R.; Legler, G.; Bourbouze, R.; Laglegerie, P. *Enzyme* **1988**, *39*, 78-89.
62. Dinur, T.; Osidcki, K.M.; Legler, G.; Gatt, S.; Desnick, R.J.; Grabowski, G.A. *Proc. Natl. Acad. Sci. USA* **1986**, *83*, 1660-1664.
63. Grabowski, G.A.; Osiecki-Newman, K.; Dinur, T.; Fabbro, D.; Legler, G.; Gatt, S.; Desnick, R.J. *J. Biol. Chem.* **1986**, *261*, 8263-8269.
64. Roeser, K.-R.; Legler, G. *Biochim. Biophys. Acta* **1981**, *657*, 321-333.
65. Legler, G.; Roeser, K.-R.; Illig, H.-K. *Eur. J. Biochem.* **1979**, *101*, 85-92.
66. Withers, S.G.; Warren, R.A.J.; Street, I.P.; Rupitz, K.; Kempton, J. B.; Aebersold, R. *J.Am. Chem. Soc.* **1990**, *112*, 5887-5889.
67. Withers, S.G.; Street, I.P. *J. Am. Chem. Soc.* **1988**, *110*, 8551-8553.
68. Withers, S.G.; Street, I.P.; Bird, P.; Dolphin, D.H. *J. Am. Chem. Soc.* **1987**, *109*, 7530-7531.
69. Withers, S.G.; Rupitz, K.; Street, I.P. *J. Biol. Chem.* **1988**, *263*, 7929-7932.
70. Fleet, G.W.J. *Tetrahedron. Lett.* **1985**, *26*, 5073-5076.
71. Post, C.B.; Karplus, M. *J. Am. Chem. Soc.* **1986**, *108*, 1317-1319.
72. Frank, R.W. *Bioorg. Chem.* **1992**, *20*, 77-88.

RECEIVED January 29, 1993

Chapter 4

A β-Glucosidase from an *Agrobacterium* sp.
Structure and Biochemistry

D. Trimbur[1], R. A. J. Warren[1], and S. G. Withers[2]

Departments of [1]Microbiology and [2]Chemistry, University of British Columbia, Vancouver, British Columbia V6T 1Z3, Canada

The ß-glucosidase (Abg) from an *Agrobacterium* sp. is a homodimer of monomers of molecular weight 51, 192. It is a retaining enzyme catalysing glycoside hydrolysis by a double-displacement mechanism in which an enzymic nucleophile (Glu358) attacks the substrate forming a glucosyl-enzyme intermediate which is then hydrolysed releasing ß-glucose. Comparisons of properties of the wild-type enzyme and its Glu358Asp mutant indicate that the principal roles for Glu358 are transition state stabilization and formation of the glycosyl-enzyme covalent intermediate, but not substrate binding. Other critical residues for enzyme activity are Gly360 and Asp374 and all three of these amino acids are conserved in the enzymes of family BGA of ß-glucosidases.

ß-Glucosidases in Bacteria

ß-Glucosidases of different specificities, such as cellobiases and aryl-ß-glucosidases, are widespread in bacteria. Cellobiases are key enzymes in the microbial degradation of cellulose, converting cellobiose, the product of the extracellular degradation of the insoluble polymer, into glucose for cell growth. Renewed interest in the bioconversion of waste cellulose-containing materials has stimulated work on ß-glucosidases. Considerable work on gene cloning and DNA sequencing has therefore led to the determination of the amino acid sequences of many ß-glucosidases from procaryotic and eucaryotic sources. The enzymes can be grouped into three families of related amino acid sequences (1), facilitating the comparison of enzymes from different sources.

0097–6156/93/0533–0042$06.00/0

Abg, a ß-Glucosidase from an *Agrobacterium* sp.

An *Agrobacterium* sp. produces an intracellular ß-glucosidase (Abg) of high specific activity and affinity for cellobiose. This enzyme hydrolyzes ß-glucosides with overall retention of anomeric configuration (2) and is thus termed a retaining enzyme. The gene for this enzyme was cloned (3) and sequenced (4) and a physical and kinetic characterization performed (5,6).

Catalytic Properties of Abg

Abg is a homodimer of monomer molecular weight 51, 192 (4,5). It has high specific activities on and high affinities for a range of ß-glycosides (Table 1) and kinetic studies (6) support a double-displacement mechanism for this enzyme involving the general acid/base catalyzed formation and hydrolysis of a glycosyl-enzyme intermediate via oxocarbonium ion-like transition states (Figure 1). 2-Deoxy-2-fluoroglucosides with reactive leaving groups, such as dinitrophenolate or fluoride, act as mechanism-based inhibitors of the enzyme (7-9). These inhibitors trap a covalent intermediate involved in the catalytic mechanism of the enzyme, thereby effectively inactivating it (Figure 2). The rationale for such trapping is that the fluorine at C-2 destabilizes the transition states for glycosyl-enzyme formation and hydrolysis, thus slowing both steps, while use of a reactive leaving group accelerates the glycosylation step sufficiently that the enzyme accumulates as its glycosyl-enzyme adduct. The destabilization of the transition states arising from fluorine substitution likely has two sources. One of these is undoubtedly inductive destabilization due to the very electronegative fluorine. The other is likely the omission of key transition state binding interactions between the 2-position and the enzyme (10,11,12).

The inactivated enzyme is still catalytically competent (13) since it reactivates spontaneously, though the rate of reactivation is very much slower than the rate of inactivation. The reactivation rate is increased dramatically by the addition of a simple ß-glucoside, such as p-nitrophenyl-ß-D-glucoside or preferably the C-linked glucoside ß-D-glucopyranosylbenzene. The rate of reactivation is dependent, in a saturable manner, on the concentration of glucosylbenzene, indicating a second binding site. Turnover of this glycosyl enzyme intermediate was subsequently shown to occur via transglycosylation, with the 2-deoxy-2-fluoro-D-glucose moiety being transferred(13) to the non-hydrolyzable glucosyl benzene to form the disaccharide derivative (Figure 2). The catalytic competence of the inactivated enzyme

Table I. Michaelis-Menten Parameters for the Hydrolysis of Aryl Glycosides by *Agrobacterium* β-Glucosidase

Phenyl glycoside substrate	$K_m(mM)$	$k_{cat}(sec-1)$	K_{cat}/K_m $(sec^{-1} mM^{-1})$
4-nitrophenyl-D-glucoside	0.078	169	2170
4-nitrophenyl-D-galactoside	5.0	275	55
4-nitrophenyl-D-mannoside	0.02	0.12	6
4-nitrophenyl-D-xyloside[a]	i) 0.22 ii) 3.3	1.85 8.0	8.4 2.4
4-nitrophenyl-D-fucoside	0.12	139	1157
4-nitrophenyl-α-L-arabinoside	0.33	32.6	98.9
2-nitrophenyl-D-galactoside	9.3	267	28.8
2-nitrophenyl-D-xyloside[a]	i) 0.025 ii) 2.4	1.04 2.36	42 1.0
phenyl-D-galactoside	16	0.91	0.06

[a]Parameters indicated are for the i) hydrolysis reaction; ii) the transglycosylation reaction.

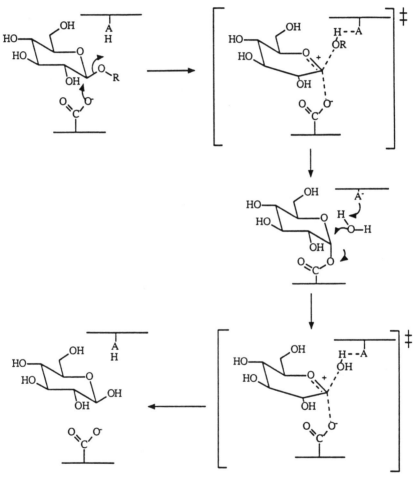

Figure 1.   Mechanism proposed for a retaining ß-glucosidase

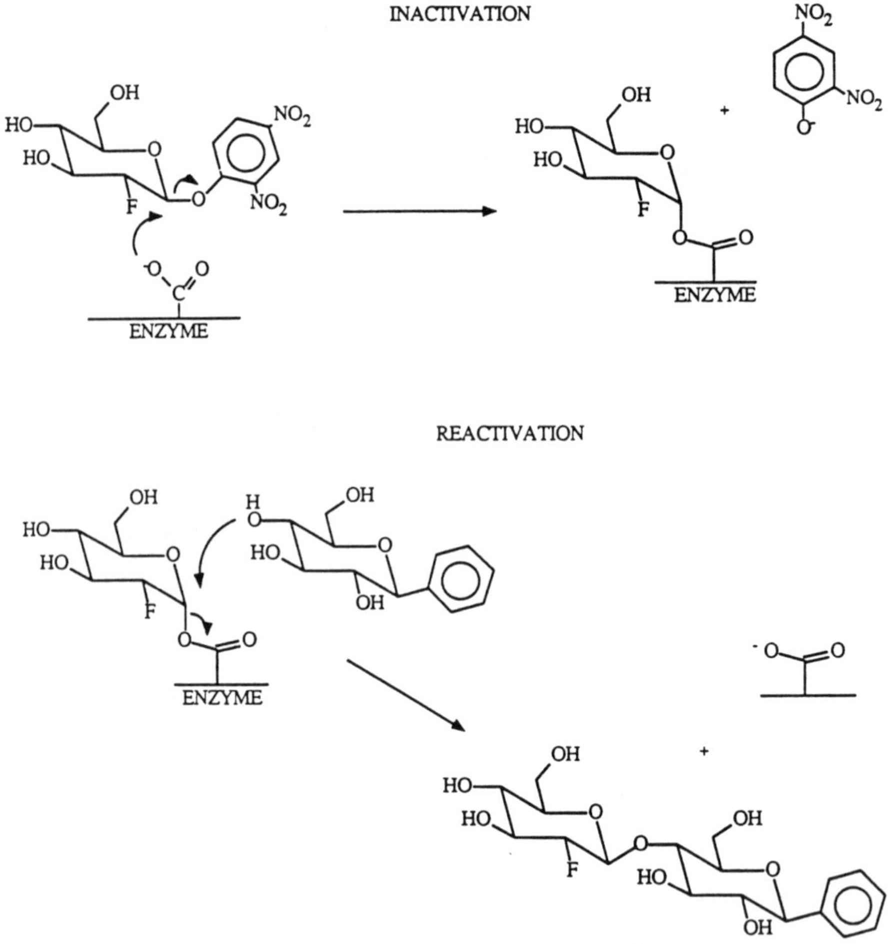

Figure 2.   Inactivation by 2-deoxy-2-fluoro glycosides and mechanism of reactivation.

indicates that the inhibitor is interacting with the normal nucleophile in the active site of the enzyme and is a true mechanism-based inhibitor, perturbing the normal mechanism by slowing the glycosylation and deglycosylation steps.

Glu358 is the Nucleophile in Abg. The nucleophile in Abg was identified as Glu358 by inactivating the enzyme with [1-^{3}H]-2′,4′-dinitrophenyl 2-deoxy-2-fluoro-ß-D-glucopyranoside, cleaving it with pepsin, and isolating and sequencing the labelled peptide (*13*). The labelled amino acid was glutamate in the sequence Y·I·T·E·N·G·A. This was the first unequivocal identification of the nucleophile in a glycosidase. The same approach was used to identify the nucleophiles in a xylanase from *Cellulomonas fimi* (*14*) and the lacZ ß-galactosidase from *Escherichia coli* (*15*), both of which are retaining enzymes.

In the families of ß-1,4-glycanases, the enzymes in a family appear to be either all retaining or all inverting (*16*). Abg is in family BGA of ß-glycosidases (*1*). Thus, since Abg is a retaining enzyme, all other enzymes in family BGA should also be retaining. The Y·I·T·E·N sequence is highly conserved in the enzymes of this family, suggesting that the glutamates corresponding to Glu358 of Abg are the nucleophiles in the other enzymes of the family (Figure 3).

Mutations around the Nucleophile of Abg. Potentially important amino acid residues in a protein can be targeted in several ways: from the three-dimensional structure of the enzyme-substrate complex or of an enzyme-substrate analogue complex; as highly conserved residues in a large family of related proteins; and by direct identification of catalytically important residues with mechanism-based inhibitors. Although nothing is known of the three-dimensional structure of Abg, the nucleophile is known and the polypeptide contains a number of highly conserved residues, including the nucleophile. Therefore, mutations were made in the nucleophile and the flanking amino acids and examined for their effects on activity (*17*).

Forty-three point mutations were generated at 22 different residues in the region surrounding the active-site nucleophile, Glu358. Only alterations at five of the residues affect Abg activity (Table 2), and four of these are conserved residues: Gly360, Asp374, R377 and Y380 (Figure 3). Replacement of Gly360 by Ser or Cys reduces activity severely. Mutations at Asp374 have very severe effects: replacement by Glu, a conservative change, reduces activity 100-fold; other changes reduce activity 2,000-fold. However, all of the mutations at Asp374 retained some measurable activity, suggesting an important but not critical role in catalysis. Since there are only five conserved Asp or Glu residues in Abg, excluding the nucleophile, it seemed possible that Asp374 serves

```
BGLS$SULSO  FFPEGLYDVLLKYWNRYH--LYMYVTENGIA-----------DDADYQRPYLISHIYQVHRAIN-S  415
BGAL$SULSO  FFPEGLYDVLLKYWNRYG--IFLVVMENGIA-----------DDADYQRPYLVSHIYQVHRALN-E  417
BGLB$ECOLI  IDPVGLRVLLNTLWDRYQK--LFFVENGLGAKDSVEADGS--IQDDYFIAYLNDHLVQVNEAIA-D  398
LACG$STAAU  IYPQGLYDQIMRVVKDYPNY-HKIYITENGLGYKDEFIES--EKTVFDDARIDYVRQHLNVIADAII-D  413
LACG$STRLC  IYPEGLYDQIMRVKNDYPNY-KKLYITENGLGYKDEFVD---NTVFDGRIDYVKQHLEVLSDAIA-D  420
LACG$LACCA  IYPRGMYDIMRIHNDYPLV-PVYVTENGIGLKESLPENATPDTVIEPKFIDYVKKYLSAMADAIH-D  416
BGLA$CLOTH  IYPEGLYDLLMLLDRDYGK--PNIVIGENGAAFKDEIGSNGK---VFDTKRIQYLKDYLTQAHRAIQ-D  391
BGLS$CALSA  VFPQGLFDLLIWIKESYPQ--IPIYITENGAAYNDIVTEDGK---VFDSKRIEYLKQHFEAARKAIE-N  400
BGLS$AGRSP  VYAPALHTLVETLYERYDLP--ECYITENGACYNMGVE-NGQ---VNIQFRLDYYAEHLGIVADLIR-D  395
BGLA$BACPO  VESRGLYEVLHYLQK-YGN--IDIYITENGACINDEVV-NGK---VQIDFRISYMQQHLVQVHRTIH-D  388
BGLB$BACPO  IHPESFYKLLTRIEKDFSKG-IPIITENGAAMRDELV-NGQ--IEDTGRHGYIEEHLKACHRFIE-E  392
LPHD4$HU    MTPFGFRRILNWLKEEY--NDFPIYVTENGVSQREE-------TDL-NDTARIYLRTYINEALKAVQ-D  406
LPHD4$RA    MTPFGFRRILNWIKEEY--NNFPIYVTENGVSHRGD-------SYL-NDTRIYLRSYINEALKAVQQD  407
LPHD3$HU    AAPWGTRRLLNWIKEEY-GDIPIYITENGVGL-TN--------PNT-EDTRIFIFYHKTYINEALKAYRLD  404
LPHD3$RA    AASFQMRRLLNWIKEEY-GDIPIYITENGVGL-TN--------PRL-ELIIEIFYKTYINEALKAYRLD  404
                                         *   *    *   *       *    *
                                         :::  :::  :::             ::  ::
                                         :::**:***                  .    .
Consensus   ..P.GL....L.......Y-----....PIYITENG.G..D.........D..RI.Y..HL......AI.-D
```

Figure 3. Alignment of residues around the nucleophile of β-glucosidases of family BGA (ref. 1) BGLS$SULSO, β-galactosidase from *Sulfolobus solfataricus*; BGALS$SULSO, β-galactosidase from *S. solfataricus*; BGLB$ECOLI, phospho-β-galactosidase from *E. coli*; LACG$STAAU, phospho-β-galactosidase from *Staphylococcus aureus*; LACG$STRLC, phospho-β-galactosidase from *Streptococcus lactis*; LACG$LACCA, phospho-β-galactosidase from *Lactobacillus casei*; BGLA$CLOTH, β-glucosidase from *Clostridium thermocellum*; BGLS$CALSA, β-glucosidase from *Caldocellum saccharolyticum*; BGLS$AGRSP, β-glucosidase from *A. faecalis*; BGLA$BACPO, β-glucosidase A from *Bacillus polymyxa*; BGLB$BACPO, β-glucosidase B from *B. polymyxa*; LPHD4$HU, LPHD4$RA, human lactase-phlorizin hydrolase domain 4; LPHD4$RA, rabbit lactase-phlorizin hydrolase domain 4; *LPHD3$HU*, human lactase-phlorizin hydrolase domain 3; *LPHD3$RA*, rabbit lactase-phlorizin hydrolase domain 3. Consensus sequence contains residues where more than 50% of the sequences have the same amino acid. Reprinted with permission from reference 17. Copyright 1992, American Society for Biochemistry and Molecular Biology, Inc.

as the acid-base catalyst which protonates the leaving group and subsequently deprotonates the water as it attacks (Figure 1). The 100-2000 fold reduction in activity upon mutation is consistent with this assignment, and also with the reduction in activity of 1000-fold or greater when the acid catalysts in hen and T4 lysozymes are mutated (18,19). Further work is required to clarify this point.

In contrast to Gly360 and Asp374, mutations at Arg377 and Tyr380 have only very small effects on enzyme activity, with only the highly disruptive change R377P decreasing activity severely. It is not at all clear why these residues are so highly conserved, since they clearly do not play an important role in catalysis. They are not involved in binding the second sugar moiety of cellobiose because the mutants have identical activities with cellobiose and p-nitrophenyl-ß-D-glucoside as substrates.

Mutation of the Nucleophile in Abg. Mutation of the nucleophile, Glu358, had much greater effects on the activity of Abg. Nine different mutations were generated at Glu358; and all of them reduced the activity at least 10,000-fold, only the conservative mutant Glu358Asp retaining measurable activity under the assay conditions (Table 2). These severe effects on enzyme activity of mutations at Glu358 are consistent with its role as the nucleophile in glycoside hydrolysis (17). The Glu358Asp mutant is especially interesting because it is the first mutant of any enzyme in which a residue which functions as an active site nucleophile has been replaced by a shorter homologue. Consequently, the mutant polypeptide has been analyzed in depth (20).

The glutamate side chain might be expected to play a minor role in substrate binding since any interactions would have to be reversed in proceeding to the transition state, thereby decreasing the amount of "differential binding energy" (21) available for selective stabilization of the transition state. Its principal roles would therefore be in electrostatic stabilization of the two positively charged transition states, and in the formation of a covalent linkage with the sugar in the glycosyl-enzyme intermediate (Figure 1). It must also function as a good leaving group for the second step of catalysis, hydrolysis of the glycosyl-enzyme. Replacement of Glu358 with Asp would be expected to have little effect on the free energy of the ground state, but it should raise the energy level of the transition states considerably. Furthermore, the covalent glycosyl-enzyme intermediate would necessarily be strained, thereby raising its free energy. The properties of the Glu358Asp mutant are consistent with this interpretation as follows.

The wild-type and the mutant enzymes have similar K_m values for substrates and similar K_i values for ground state inhibitors, indicative

Table II. Mutations Generated in Active Site Region

Strain	Mutation	pNPG Activity*
ABG (wild-type)		1.9
1068AB78	H 339 L	1.8
1068AB59	H 339 P	2.0
1068AB104	E 343 V	1.7
1068AW101-1	T 344 S	2.1
1068AB100	Y 346 F	1.4
1068AB9	D 350 V	3.0
1068AB174	Y 355 F	3.4
1091GP3-1	T 357 P	2.2
E358D-1	E 358 D	0.0008
E358N-1	E 358 N	0.0002
E358Q-1	E 358 Q	0.0002
E358G-1	E 358 G	0.0001
E358C-8	E 358 C	0.0001
E358H-4	E 358 H	0.0001
1068AW116	E 358 V	0.0001
1068AW88	E 358 A	0.0002
1091GW5	E 358 K	0.0001
1068AB200	N 359 S	0.6
1068GW69	G 360 C	0.0001
1091GW2	G 360 S	0.002
1068AW28	C 362 Y	0.0001
1068AW92	Y 363 S	3.2
1068AB184	Y 363 F	3.8
1068GB43	M 365 I	1.5
1086GB8	V 367 I	2.8
1068AB192	E 368 D	1.6
1068GB9	G 370 C	1.9
1068GB52	G 370 S	1.9
1068GB1	E 371 K	1.4
D374-2	D 374 V	0.001
D374-6	D 374 G	0.003
1091GW6	D 374 N	0.006
1068AW117	D 374 A	0.007
D374-5	D 374 E	0.02
1068AW6	R 377 P	0.002
R3771-19	R 377 I	0.1
R377T-13	R 377 T	0.3
R377-42	R 377 K	0.8
Y380-31	Y 380 C	0.1
Y380-30	Y 380 S	0.3
Y380-42	Y 380 F	0.8

*μmol of pNP released/min/mg of total cell protein

of a minimal role of Glu358 in substrate binding (Tables 3, 4). However, the activity (k_{cat}/K_m) of the mutant is much less than that of the wild-type enzyme, and its binding of transition state analogues is much weaker, indicative of a role of Glu358 in transition state stabilization. An estimate of the extent to which each transition state is destabilized in the mutant is obtained by calculating the reductions in the rates of each individual step. The reduction in the rate of the glycosylation step in the mutant is evident from the hydrolysis of m-nitrophenyl-ß-D-glucoside, a substrate for which the glycosylation step is rate-determining in both enzymes: k_{cat} for the mutant is reduced 2,500-fold, corresponding to an increase in $\Delta G^{\ddagger}$ of about 4.5 kcal mol^{-1}. Similarly, the mutant forms the glycosyl-enzyme intermediate from 2',4'-dinitrophenyl 2-deoxy-2-fluoro-ß-D-glucopyranoside almost 6,000-fold more slowly than the wild-type enzyme. Comparison of the deglycosylation rate constants is difficult because a substrate has not been identified for which the deglycosylation step is rate-limiting with the mutant enzyme. However, the k_{cat} values for the hydrolysis of the 2',5'-dinitrophenyl-ß-D-glucoside by the mutant and wild-type enzymes (1.22 sec^{-1} and 150 sec^{-1}, respectively) suggest that the deglycosylation step is slowed _less_ than 125-fold in the mutant. Furthermore, hydrolysis of the 2-deoxy-2-fluoro-∝-D-glucopyranosyl-enzyme intermediate is some 1,400-fold _faster_ for the mutant enzyme than for wild type, indicative of very substantial destabilization of the intermediate in the mutant. A stretched transition state in the mutant enzyme is indicated by the secondary deuterium kinetic isotope effect (k_H/k_D = 1.17) measured for the hydrolysis of 2',4'-dinitrophenyl-ß-D-glucoside. The large value for this effect reflects the extensive oxocarbonium ion character of the transition state in the mutant enzyme, even though the stabilizing negative charge is further away than in the wild-type enzyme. This strongly suggests that reaction of substrate with the wild-type enzyme is highly pre-associative, which decreases the observed isotope effect. This should be viewed in the light of a recent proposal that reactions of anionic nucleophiles at acetal centres will tend to be pre-associative (22).

Despite the consistency of the results, it must be borne in mind that the Glu358Asp mutation may cause significant structural changes in the enzyme. Clearly, determination of the structure of Abg is crucial, and crystallization of the enzyme is well in hand.

Other ß-Glucosidases of Family BGA

Family BGA of ß-glucosidases contains enzymes from procaryotes, eucaryotes and archaebacteria (1). Since Glu358 is conserved in all of

Table III. Kinetic Parameters for Aryl Glucosides with the Asp358 Mutant and Wild Type Enzymes

Substrate (Aryl β-glucoside)	Aglycone pK_a	k_{cat} (sec^{-1}) E358D	K_m (mM) E358D	k_{cat} (sec^{-1}) WILD TYPE	K_m (mM) WILD TYPE
2',4'-dinitrophenyl-	3.96	0.72	0.054	87.9	0.031
2',5'-dinitrophenyl-	5.15	1.22	0.20	120	0.045
3',4'-dinitrophenyl-	5.35	0.040	0.068	185	0.033
2'-nitro,4'-chlorophenyl-	6.45	0.054	0.07	144	0.013
4'-nitrophenyl-	7.18	0.021	0.093	169	0.078
3'-nitrophenyl-	8.39	0.043	1.9	108	0.19

Table IV. Inhibition Constants for Native and Glu358Asp β-Glucosidase

Inhibitor	Category	K_i-Native[1] (mM)	K_i - Mutant (mM)	K_i Ratio (Mutant/Native)
β-Glucosyl benzene	Ground state	3.4	17	5
β-Glucosylamine	Ground state/ Transition state	0.16	8.2	51
Gluconolactone	Transition state	0.0014	0.53	380
Gluconophenylurethane	Transition state	0.0012	0.70	580
Gluconohydroximolactone	Transition state	0.03	3.6	120

Data from Kempton & Withers (1992)

them, it seems likely that it is the nucleophile in all of them. If so, it should be targeted by the mechanism-based inhibitors used for Abg; and its mutation to Asp should have much the same effects as the Glu358 Asp mutation in Abg. The acid-base catalytic residue is probably identical in all of the enzymes. We are analyzing mutations in other conserved residues in Abg in an effort to identify the acid-base catalyst and to analyse its role in the enzyme reaction.

Acknowledgment

This research was supported by the Natural Sciences and Engineering Research Council of Canada.

Literature cited

1. Henrissat, B.M. *Biochem. J.*, **1991**, *280*, 309-316.
2. Withers, S.G.; Dombroski, D.; Berven, L., Kilburn; D.G., Miller, R.C., Jr.; Warren, R.A.J.; Gilkes, N.R. *Biochem. Biophys. Res. Commun.* **1986**, *139*, 487- 494.
3. Wakarchuk, W.W.; Kilburn, D.G.; Miller, R.C., Jr.; Warren, R.A.J. *Mol. Gen. Genet.* **1986**, *205*, 146-152.
4. Wakarchuk, W.W.; Greenberg, N.M.; Kilburn, D.G.; Miller, R.C., Jr.; Warren, R.A.J. *J. Bacteriol.* **1988**, *170*, 301-307.
5. Day, A.G.; Withers, S.G. *Biochem. Cell. Biol.* **1986**, *64*, 914-922.
6. Kempton, J.B.; Withers, S.G. *Biochemistry* **1992**, *31*, 9961-9969.
7. Withers, S.G.; Street, I.P.; Bird, P.; Dolphin, D.H. *J. Am. Chem. Soc.* **1987**, *109*, 7530-7531.
8. Withers, S.G.; Street, I.P. *J. Am. Chem. Soc.* **1988**, *110*, 8551-8553.
9. Withers, S.G.; Rupitz, K.; Street, I.P. *J. Biol. Chem.* **1988**, *263*, 7929-7932.
10. Street, I.P.; Kempton, J.B.; Withers, S.G. *Biochemistry* **1992**, *31*, 9970-9978.
11. McCarter, J.; Adam, M.; Withers, S.G. *Biochem. J.* **1992**, *286*, 721-727.
12. Konstantinidis, A.; Sinnott, M.L. *Biochem. J.* **1991**, *279*, 587-593.
13. Withers, S.G.; Warren, R.A.J.; Street, I.P.; Rupitz, K.; Kempton, J.B.; Aebersold, R. *J. Am. Chem. Soc.* **1990**, *112*, 5887-5889.
14. Tull, D.; Withers, S.G.; Gilkes, N.R.; Kilburn, D.G.; Warren, R.A.J.; Aebersold, R. *J. Biol. Chem.* **1991**, *266*, 15621-15625.
15. Gebler, J.C.; Aebersold, R.; Withers, S.G. *J. Biol. Chem.* **1992**, *267*, 11126-11130.
16. Gebler, J.; Gilkes, N.R.; Claeyssens, M.; Wilson, D.B.; Béguin, P.; Wakarchuk, W.W.; Kilburn, D.G.; Miller, R.C., Jr.; Warren, R.A.J.; Withers, S.G. *J. Biol. Chem.* **1992**, *267*, 12559-12562.

17. Trimbur, D.E.; Warren, R.A.J.; Withers, S.G. *J. Biol. Chem.* **1992**, *267*, 10248-10251.
18. Anand, N.N.; Stephen, E.R.; Narang, S.A. *Biochem. Biophys. Res. Commun.* **1988**, *153*, 862-868.
19. Malcolm, B.A.; Rosenberg, S.; Corey, M.J.; Allen, J.S.; De Baetselier, A.; Kirsch, J.F. *Proc. Natl. Acad. Sci. U.S.A.* **1989**, *86*, 133-137.
20. Withers, S.G.; Rupitz, K.; Trimbur, D.; Warren, R.A.J. *Biochemistry* **1992**, *31*, 9979-9985.
21. Fersht, A.R. Enzyme Structure and Mechanism. 1985, **2nd Edition**, Freeman, New York.
22. Banait, N.S., Jencks, W.P. *J. Amer. Chem. Soc.* **1991**, *113*, 7958-7966.

RECEIVED February 1, 1993

Chapter 5

Deletion of the *Trichoderma reesei* β-Glucosidase Gene, *bgl1*

Tim Fowler

Genencor International Inc., 180 Kimball Way,
South San Francisco, CA 94080

A targeted gene disruption technique has been used to remove the *bgl1* gene encoding an extracellular ß-glucosidase from the genome of the cellulolytic fungus *Trichoderma reesei*. ß-glucosidase null strains have been used to investigate its role as a component of the cellulase enzyme system in the hydrolysis of cellulose and induction of the other cellulolytic enzyme components.
  Compared to the parent strains ß-glucosidase null mutants have the following properties when using lactose and Avicel as carbon sources; 1) growth was unaffected, 2) levels of extracellular protein and total endoglucanase production were seen to temporally lag those levels observed in the parent strain, and 3) the mRNA levels of the other cellulase genes showed a corresponding lag in induction indicating that the absence of extracellular ß-glucosidase has an effect on the co-ordinate regulation of the other cellulase genes at the level of transcription. These results, taken together, indicate that extracellular ß-glucosidase is required for rapid induction of the other cellulase enzymes in *T. reesei*. These data are compared to those from similar experiments performed on a *T. reesei* strain that had been deleted for the gene encoding cellobiohydrolase II, *cbh2*.

This chapter describes the cellulase enzyme system of *Trichoderma reesei* and in particular the genetic manipulation of this complex to produce novel strains in which the cellulase enzyme profile has been altered. It focuses is on how the cellulase genes are regulated in the absence of the ß-glucosidase (*bgl1*) and cellobiohydrolase II (*cbh2*) genes.
  Cellulase, from the filamentous fungus *Trichoderma reesei,* was first described in the 1950's by Reese and Mandels. Since then, the cellulase enzyme complex of *T. reesei* has been shown to consist of at least three types of enzymes that together convert native crystalline cellulose to oligosaccharides and glucose (*1*). The endoglucanases (1,4-ß-D-glucan 4-glucanohydrolase, EC 3.2.1.4) and cellobiohydrolases (1,4-ß-D-glucan cellobiohydrolase, EC 3.2.1.91) are thought to act in synergy to hydrolyze cellulose into small cellooligosaccharides. The latter (mainly cellobiose) are

subsequently hydrolyzed to glucose by ß-glucosidase (ß-D-glucoside glucohydrolase, EC 3.2.1.21).

The role of the individual cellulase components and in particular the extracellular ß-glucosidase in cellulose hydrolysis and in the regulation of the cellulase enzyme system is currently under investigation at Genencor International. The exact biochemistry and synergy between the different cellulases of *T. reesei* is the subject of a great deal of investigation and has not been fully resolved, however, the genes encoding cellobiohydrolase I (*cbh1*) (23) cellobiohydrolase II (*cbh2*) (24,25) endoglucanase I (*egl1*) (26) and endoglucanase II (*egl3*) (27) have been cloned and sequenced. Thus we are presented with an opportunity to look in more detail at the cellulase enzyme system by manipulating each individual component at the genetic level. We have previously reported the molecular cloning and sequence analysis of the extracellular ß-glucosidase gene, *bgl1* from *T. reesei*. (6).

Methods for transformation and genetic manipulation have been recently been made available to *Trichoderma reesei* (2) and have been employed to manipulate the cellulase enzymes of strains resulting in the complete removal or overproduction of either individual or multiple cellulase components (3, 4, 5, 6, 7). This is industrially very important as it means strains with novel cellulase profiles can be genetically tailored to defined applications (See Table I from ref. 8). Historically this alteration of cellulase component ratios was achieved by blending with other cellulase preparations (28, 29, 30) or by variation in fermentation conditions. In addition, it has been possible to look more closely at the role of individual cellulase components in the biochemistry and synergy of cellulose hydrolysis as well as investigating how the cellulase genes are coordinately regulated upon growth of the fungus on cellulose.

The role of the cellulase gene products, including ß-glucosidase, in cellulose hydrolysis and in the regulation of the cellulase enzyme system is currently under investigation at several laboratories. Attempts have been made to obtain ß-glucosidase deficient strains of *T. reesei* by mutagenesis with either UV light (9) or gamma irradiation (10). Their inability to obtain non-leaky ß-glucosidase null mutations suggested to the authors that either ß-glucosidase is an essential gene (for example, involved in morphogenesis (11) or as an essential structural component of the cell wall) or there are at least two different ß-glucosidase genes in *T. reesei*. (12). Presumably one function of ß-glucosidase is the breakdown of cellobiose, produced by the cellobiohydrolases and endoglucanases, to provide glucose as a carbon source. Expression of the cellulase genes is tightly regulated, being repressed in the presence of glucose and induced in the presence of cellulose as a sole carbon source. However since glucose results from the breakdown of a largely insoluble polymeric substrate a mechanism presumably exists to release soluble oligosaccharides from cellulose which can mediate the induction of the cellulase genes. A low constitutive level of cellulase expression has been postulated (13, 14, 15) which could presumably result in low levels of cellobiose in the presence of cellulose. Another function of ß-glucosidase may be to use the resultant cellobiose and glucose (via a transglycosylation reaction) to produce oligosaccharides that have been shown to act as potent inducers of the cellulase enzyme system (1, 16, 17).

Here we describe the deletion of the *bgl1* gene from *T. reesei* strain RL-P37 (18) and the use of this well defined null-mutation as a first step in the investigation of the role of ß-glucosidase in cellulase gene induction.

Deletion of the *bgl1* Gene from *T. reesei*.

Mutants of *Trichoderma reesei* lacking the coding sequence for the extracellular ß-glucosidase gene, *bgl1*, were obtained by a targeted gene replacement event (19). A gene replacement vector was first constructed (Illustrated in Figure 1, Ref. 19). The vector was digested with *Hind*III to release a linear fragment in which the *bgl1* coding sequences were replaced with the *pyrG* gene from *Aspergillus niger* (*pyrG* encodes an

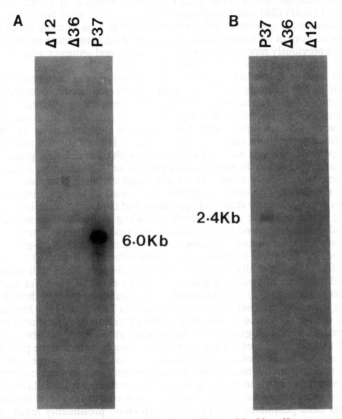

Figure 1. Southern and Northern blot analysis of *bgl1* null mutants.   Panel A.
Southern analysis of genomic DNA extracted from strains Δ12, Δ36, and RL-P37.
10 μg DNA was digested with *Hind*III prior to electrophoresis and blotting onto
Nytran.   The 2290 bp *Apa*I/*Eco*RV restriction fragment containing the *bgl1*
coding sequences was used as a probe (6). Panel B. Northern analysis of 20 μg
total mRNA isolated from RL-P37, Δ36 and Δ12.   The same radiolabelled
*Apa*I/*Eco*RV *bgl1* gene fragment was used as a probe.
Source: Reproduced with permission from Figure 2 ref. 19.

enzyme required in the uridine biosynthetic pathway). Purified linear *Hind*III fragment from the disruption vector was used to transform a *T. reesei* strain RL-P37 *pyrG69*, a uridine requiring auxotroph, to prototrophy. Transformants lacking the *bgl1* coding sequences were screened initially by dot blot hybridization using *bgl1* coding sequences as a probe. Transformants identified in this way were subjected to Southern and Northern blot analysis (Figure 1). Transformant genomic DNA was subjected to digestion with *Hind*III prior to gel electrophoresis and blotting. A probe consisting of the coding region of the *bgl1* gene was used to determine the integrity of the *bgl1* locus. The parental strain, RL-P37 *pyrG69*, gave the expected 6.0 Kb fragment compared to two strains deleted for *bgl1*, Δ12 and Δ36, which showed no hybridization (Panel A). To confirm that transcription from the *bgl1* gene had also been disrupted, total RNA was isolated from the deletion transformant and RL-P37 *pyrG69*. Northern blot analysis using the same *bgl1* probe indicated that *bgl1* specific mRNA present in RL-P37 *pyrG69* is absent in the *bgl1* deletion strains (Panel B). Western blot analysis established that extracellular ß-glucosidase, encoded by the *bgl1* gene, was also absent in these strains (See Figure 3 Ref. *19*)

Evidently the extracellular ß-glucosidase encoded by the *bgl1* gene is not essential as demonstrated by the successful selection of a disruption within *bgl1*. Indeed Δ*bgl1* strains sporulate normally and will grow on cellobiose and cellulose as a sole carbon source. Moreover, on the basis of DNA hybridization experiments we have not seen any evidence of sequences showing even weak homology to the *bgl1* gene within the genome of *T. reesei*. If there are other ß-glucosidase enzymes produced by *T. reesei* (perhaps responsible for a low level of residual activity (0.302 µmol PNP/mg/min for RL-P37 and 0.021 for Δ36)) they must be genetically distinct from *bgl1*. When supernatant from the *bgl1* disruption strain was subjected to Western blot analysis using antibodies raised against the extracellular ß-glucosidase no signal was observed. This argues against other ß-glucosidase enzyme forms arising from post-transcriptional modification of the *bgl1* mRNA or post-translational modification of the ß-glucosidase protein.

Protein Export and Cellulase Production from *T. reesei* Deficient in Extracellular ß-Glucosidase.

The deletion of the *bgl1* gene from *T. reesei* provides a host with which to study the role of extracellular ß-glucosidase in the regulation of other enzymes of the cellulase mixture and its function in the hydrolysis of cellulose. The effect of the absence of the *bgl1* gene product on cell growth, cellulase induction and extracellular protein production was determined by assaying culture supernatants following growth on a variety of carbon sources.

Data are presented more fully in Fowler & Brown (*19*). The effect of disruption of *bgl1* on cell growth was first determined. Growth of RL-P37 and Δ36 were seen to be essentially identical in conditions that favored either the induction (lactose and cellobiose as a carbon source) or non-induction (glucose as a carbon source) of the cellulases over a period of 4 days (See Figure 4. in Ref. *19*). Growth on Avicel was observed to be similar between the strains but was not measured *per se* with this insoluble substrate due to the difficulty of measuring mycelial dry weight in the presence of residual Avicel. These experiments indicate that any changes measured in extracellular protein levels, cannot be attributed to differential cell growth following experiments on soluble substrates.

In this regard lactose is an ideal substrate as it allows for induction of the cellulase enzymes in a manner analogous to that observed for crystalline cellulose (Avicel) while at the same time avoids the problems of quantitation associated with an insoluble substrate. In the case of a deletion in the *bgl1* gene a lag in the production of cellulase enzymes was observed even though growth of the deletion strain, when compared to the parental strain, was the same (*19*).

An experimental comparison was made to strains deleted for the gene encoding cellobiohydrolase II. This enzyme has been reported to be involved in cellulase induction (20, 21, 22). A *cbh2* deletion vector and RL-P37 strains deleted for the *cbh2* gene were obtained essentially as described for the *bgl1* gene (6, M. Ward unpublished results). Figure 2 illustrates that a *T. reesei* strain deleted for the *cbh2* gene displays normal growth on Vogels + 1% lactose.

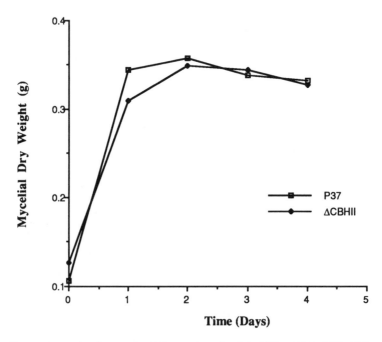

Figure 2. Comparison of cell growth of *T. reesei* strains RL-P37 and RL-P37 *cbh2⁻* when grown on Vogels + 1% lactose medium.

These results showing similar growth rates on lactose for the ΔCBHII strain and the parental strain, RL-P37, are in agreement with those described previously (20). Seiboth et al (20) noted, in contrast to that seen with a ß-glucosidase deleted strain, that a *cbh2⁻* strain grown on Avicel cellulose displayed a lag in growth of 18-24 hours. It is possible that this delay in growth and corresponding delay in the induction of cellulase is as a result of a reduced ability to utilize Avicel as a carbon source.

Transcriptional Regulation of the Cellulase Genes in the Absence of *bgl1*.

To determine whether ß-glucosidase exerts an effect (directly or indirectly) on transcriptional regulation of the other cellulase genes, quantitative Northern blot experiments were performed. Results are presented more fully in Fowler & Brown (19). Total mRNA was prepared from *T. reesei* strains RL-P37 and Δ36 over a period of 3 days following induction on Avicel. Equivalent quantities of mRNA (as determined spectrophotometrically) were fractionated by formaldehyde gel electrophoresis, transferred to Nytran, and hybridized to radiolabelled probes consisting of isolated gene fragments from the *cbh1* (23), *cbh2* (24, 25), *egl1* (26), and *egl3* (27) genes. The results for the Northern probed with *cbh1* and quantitated are given in Figure 3.

The results were essentially the same for all of the other cellulase genes probed (See Figure 5 Ref. *19*). That is, rapid co-ordinate induction of all the cellulase genes was observed after 24 hours in RL-P37; thereafter message levels are seen to decline. In the absence of ß-glucosidase a 12-36 hour lag in cellulase gene transcription is observed before message levels are seen to rise to levels comparable to wild type expression.

The lag in cellulase gene mRNA and protein production by strain Δ36 is abolished if the same experiment is repeated in the presence of 1 mM sophorose (See Figure 6 Ref. *19*). Thus a known inducer of the *T. reesei* cellulase system (sophorose) can replace the function of ß-glucosidase in rapid induction of the other genes of the cellulase system. These data indicate that some basic cellular function affecting the intrinsic capability of the cell to produce mRNA has not been influenced as a result of the ß-glucosidase deletion. In addition, it can be inferred that a lack of ß-glucosidase does not impair secretion of the other cellulase enzymes.

In order to determine the effect of a deletion in the *cbh2* gene on *cbh1* transcription these experiments were repeated and mRNA was isolated from *T. reesei* strains RL-P37 and RL-P37 *cbh2* ⁻ over a period of 3 days following induction on lactose. Lactose was chosen as a carbon source to circumvent any differences that may be observed as a result of differential growth rates on Avicel. It is important to note that if cellulase mRNA levels are analyzed from mRNA isolated from RL-P37 Δ36 a lag in *cbh1* mRNA production is also observed when grown on lactose as a carbon source. The results are shown in Figure 4.

The *cbh1* message level peaks at the same time in the *cbh2*⁻ strain as in RL-P37 but does not reach the same maximal value. It is not clear if this reduction is biologically significant and represents a reduced *cbh1* transcription rate, reduced message stability or a subtle feature of the utilization of the carbon source, lactose.

Thus preliminary studies indicate that in the absence of ß-glucosidase, induction of the other cellulase genes is delayed at the level of transcription. No such lag is observed for *cbh1* gene message in a *cbh2* ⁻ strain. It is interesting to note that the *bgll* deficient strain is able to utilize cellobiose as a carbon source as efficiently as the parental strain RL-P37 indicating the presence of alternative uptake and utilization mechanisms.

Conclusions and Future Directions.

This chapter describes the deletion of the *bgll* gene encoding the extracellular ß-glucosidase from *T. reesei* along with experiments which indicate that the *bgll* gene product is non essential but its presence results in a more rapid induction of the cellulase enzyme system. The observation that induction of the cellulase enzymes occurs, albeit delayed, in the absence of *bgll* gene product suggests that other processes are apparently capable of substituting for the proposed inducer formation function of ß-glucosidase. However, this substitute function occurs with a reduced efficiency resulting in a lag in production of cellulase mRNA and a more moderate decline in mRNA levels. It is possible that other cellulase genes play a role in the mechanism that "senses" the presence of cellulose in the surrounding environment resulting in the efficient induction of the cellulase enzymes of *T. reesei*. CBHII has been implicated as having a role in cellulase gene regulation (20, 21, 22). Our results suggest that cellulase gene mRNA levels are affected upon disruption of the *cbh2* gene however the timing of mRNA induction remains the same. It remains to be seen to what extent (if any) the other cellulase enzymes are involved in the process of

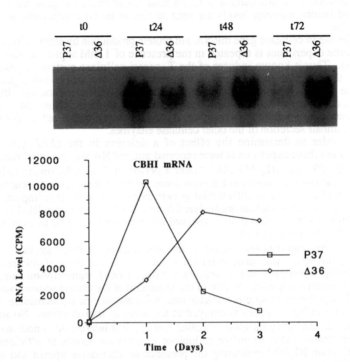

Figure 3. Quantitative mRNA analysis of *cbh1* cellulase gene expression in *T. reesei* strains RL-P37 and Δ36. Strains were grown in Vogels medium with 1% (w/v) Avicel. 20 μg total mRNA was loaded in each lane, electrophoresed and blotted onto Nytran. Bands of hybridization were visualized on X-ray film and then quantitated using the Ambis Imaging System (Ambis Corp., San Diego, CA). Note that small errors in the spectrophotometric quantitation of the mRNA may exist.
Source: Adapted with permission from Figure 5 ref. 19.

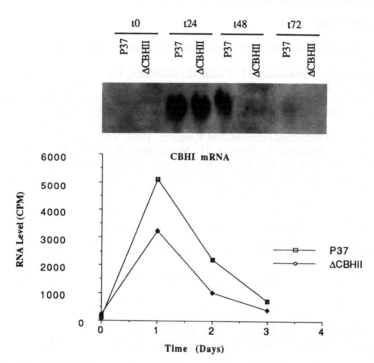

Figure 4. Quantitative mRNA analysis of *cbh1* cellulase gene expression in *T. reesei* strains RL-P37 and RL-P37ΔCBHII.   Strains were grown in Vogels medium with 1% (w/v) Lactose. 20 μg total mRNA was loaded in each lane, electrophoresed and blotted onto Nytran. Note that small errors in the spectrophotometric quantitation of the mRNA may exist. Radiolabelled DNA from the coding region of the *cbh1* gene was used as a probe.

induction. We are continuing to use targeted gene disruption as a tool in the analysis of cellulase gene regulation and function in *Trichoderma reesei*.

Acknowledgments

The author would like to thank Drs. Mick Ward and, Ross D. Brown Jr. for the many thoughtful discussions, advice and critical reading of the manuscript.

Literature Cited

1. Gritzali, M.; Brown, R.D., Jr. *Advances in Chemistry series*. **1979**, *181*, 237-260.
2. Pentitila, M. ; Nevalainen, H.; Ratto, M.; Salminen, E.; Knowles, J. *Gene* **1987**, *61*, 155-164.
3. Knowles, J.; Penttila, M.; Harkki, A.; Nevalainen, H.; Teeri, T.; Saloheimo, M.; Uusitalo, J. In; Molecular biology of Filamentous Fungi. Eds. Nevalainen, H. and Penttila, M., Proceedings of the EMBO-ALKO Workshop. Foundation for Biotechnical and Industrial Fermentation Research, 1989, Vol 6; pp 113-118.
4. Harkki, A., Mantyla, A., Penttila, M., Muttilainen, S., Buhler, R., Suominen, P., Knowles, J., and Nevailainen, H. *Enzyme Microb. Technol.* **1991** *13*, 227-233.
5. Nevalainen K.M.H.; Penttilla M.E.; Harkki A.; Teeri T.T.; Knowles J. In; Molecular Industrial Mycology. Eds. Leong S.A. and Berka R.M., Publ. Marcel Dekker, Inc. New York, NY, 1991, Vol 8; pp 129-148.
6. Barnett, C.B.; Berka R.; Fowler T. *Bio/technology* **1991**, *9*, 562-567.
7. Teeri, T.T.; Penttila, M.; Keranen, S.; Nevalainen, H.; and Knowles, J. In; Biotechnology of Filamentous Fungi, Technology and Products. Eds. Finkelstein D.B. and Ball C., Publ. Butterworth-Heinemann, Boston, MA, 1992, Vol 21; pp 417-445.
8. Uusitalo, J.M.; Nevalainen, K.M.H.; Harkki, A.M.; Knowles, J.K.C.; Penttila M.E. *J. Biotechnol.* **1991**, *17*, 35-50.
9. Mishra, S.; Rao, S.; Deb, J.K. *J. Gen. Micro.* **1989**, *135*, 3459-3465.
10. Strauss, J.; Kubicek, C.P. *J. Gen. Micro.* **1990**, *136*, 1321-1326.
11. Jackson, M.A.; Talburt, D.E.   *Exp. Micol.* **1988**, *12*, 203-216.
12. Chen, H.; Hayn, M.; Esterbaur, H. *Biochim. Biophys. Acta.* **1992**, *1121*, 54-60.
13. Mandels, M. Parrish, F.W., and Reese, E.T.   *J. Bacteriol.* **1961**, *83*, 400-408.
14. Sternberg D. and Mandels G.R. *J. Bacteriol.* **1979** *139*,761-769.
15. El Gogary S., Leite A., Crivellaro O., Eveleigh D.E., and El Dorry H. *Proc. Nat. Acad. Sci.* **1989** *86*, 6138-6141.
16. Valheri, M.P.; Leisola, M.; Kaupinnen, V. *Biotechnol. Lett.* **1979**, *1*, 41-46.
17. Kubicek, C.K. *J. Gen. Microbiol.* **1987**, *133*, 1481-1487.
18. Sheir-Ness, G. and Montenecourt, B.S. *Applied Microb. and Biotechnol.* **1984**, *20*, 46-53.
19. Fowler, T.; Brown, R.D. Jr. *Mol. Micro.* **1992**, *6*, 3225-3235.
20. Seiboth, B.; Messner, R.; Gruber, F.; Kubicek, C.P. *J. Gen. Micro.* **1992**, *138*, 1259-1264
21. Kubicek-Pranz, E.M.; Gruber, F.; Kubicek, C.P. *J. Biotechnol.* **1991**, *20*, 83-94.
22. Messner, R.; Kubicek-Pranz, E.M.; Gsur, A.; Kubicek, C.P. *Arch. Micro.* **1991**, *155*, 601-606.
23. Shoemaker, S.; Schweickart, V.; Ladner, M.; Gelfand, D.; Kwok, S.; Myambo, K.; Innis, M. *Bio/Technology* **1983**, *1*, 691-696.
24. Chen, M.C.; Gritzali, M.; Stafford, W.D. *Bio/Technology* **1987**, *5*, 274-278.
25. Teeri, T. T.; Lehtovaara, P.; Kauppinen, S.; Salovuori, I.; Knowles, J.K.C. *Gene* **1987**, *51*, 43-52

26. Penttila, M.; Lehtovaara, P.; Nevalainen, H.; Bhikhabhai, R.; Knowles, J.K.C.   *Gene* **1986**, *45*, 253-263.
27. Saloheimo, M.; Lehtovaara, P.; Penttila, M.; Teeri, T.T.; Stahlberg, J.; Pettersson, G.; Claeyssens, M.; Tomme, P.; Knowles J.K.C.   *Gene* **1988**, *63*, 11-21.
28. Enari T.M.; Niku-Paavola M.L.; Harju L.; Lappalainen A.; Nummi M. *J. Appl. Biochem.* **1981**, *3*, 157-163
29. Sternberg D.; Vijayakumar P.; Reese E. T. *Can. J. Microbiol.* **1977**, *23*, 139-147.
30. Kadam S. K.; Demain A. L. *Biochem. Biophys. Res. Comm.* **1989**, *161*, 706-711

RECEIVED January 12, 1993

Chapter 6

Molecular Biology and Enzymology of Human Acid β-Glucosidase

Gregory A. Grabowski[1,2], Anat Berg-Fussman[2], and Marie Grace[2]

[1]Department of Pediatrics, Division of Human Genetics, Children's
Hospital Medical Center, Cincinnati, OH 45229
[2]Department of Pediatrics, Division of Medical and Molecular Genetics,
Mount Sinai Medical Center, New York, NY 10029

Acid β-glucosidase, a membrane bound lysosomal
hydrolase, is critical to glycosphingolipid metabolism.
Gaucher disease results from a variety of mutations at
this locus. Kinetic studies with activators and inhibitors
coupled with site-directed mutagenesis have provided
insight into structure/function relationships and this
enzyme's mechanism of reaction. Such studies suggest
a discrete organization of functions or properties that
may have implications for understanding the molecular
pathology of Gaucher disease and for design of
enzymes or genes for its therapy.

Glucosylceramide is the penultimate intermediate in the catabolic pathway of most
complex glycosphingolipids. Cleavage of the glucosidic bond by acid β-glucosidase
(E.C. 3.2.1.45; N-acyl-sphingosyl-β-$\underline{D}$-glucopyranoside; glucohydrolase) produces
β-glucose and ceramide, the latter being further degraded by acid ceramidase to
sphingosine and fatty acid (1,2). Major interest in this catabolic step derives from
the occurrence of a human disease due to inherited defects in this hydrolytic
activity. The eponym "Gaucher disease" designates the heterogeneous sets of signs
and symptoms in patients with such defective intracellular hydrolysis of
glucosylceramide and related glucosphingolipids. At the biochemical and molecular
levels, these autosomal recessively inherited disorders most commonly result from
mutations at the locus on chromosome 1 that encodes the lysosomal hydrolase, acid
β-glucosidase or glucocerebrosidase (3). Rare examples of mutations at the
prosaposin or "activator" locus on chromosome 10 also lead to severe phenotypes
with glucosylceramide accumulation. An initial nosology includes the three major
types of Gaucher disease that have been delineated by the absence (type 1) or
presence and severity (types 2 and 3) of primary central nervous system
involvement (Table 1). Recent progress in elucidating the etiology of the Gaucher

0097–6156/93/0533–0066$06.00/0

disease variants has provided insights into the molecular biology and enzymology of human acid β-glucosidase and its "activator" proteins.

TABLE 1: GAUCHER DISEASE -- CLINICAL TYPES

Clinical Features	Type 1	Type 2	Type 3
Onset	Childhood/ Adulthood	Infancy	Juvenile
Hepatosplenomegaly	+ → +++	+++	+++
Hypersplenism	+ → +++	++	++
Bone Degeneration	- → +++	--	+ → +++
Neurodegeneration	Absent	++++	++ → +++
Death	Childhood/ Adulthood	By 2 years	2nd → 4th Decades
Ethnic Predilection	Ashkenazi Jews	None	Swedish

Molecular Biology of Acid β-Glucosidase and Prosaposin
    Molecular Biology of Acid β-Glucosidase. The complete sequences of the human acid β-glucosidase structural gene (7604 bp) and its unprocessed pseudogene have been reported (*4*). Both human sequences are contained within a single 32 kb fragment from human chromosome 1(q21→31) with the pseudogene being 3' to the structural gene (*4-6*). In the regions present in both sequences, 96% nucleotide identity was found. The major differences between the pseudogene and structural gene are the presence of four large intronic deletions in the pseudogene. These deletions represent regions of *Alu* sequences flanked by direct repeats within the structural gene. In addition, the pseudogene contains numerous point substitutions and small deletions in introns and exons scattered throughout the sequence. The promoter region of the structural gene contains two TATA boxes and two possible CAAT-like boxes about 200-300 bp upstream from the ATG initiation codon (*4,7,8*). These results are interesting since promoter elements containing TATA and CAAT boxes have been identified most commonly in highly regulated genes (*9*). In transfection experiments with pSVOCAT constructs, the promoter sequences from the structural gene were at least eight times stronger than the corresponding sequences from the psuedogene (*7,8*)
    The mouse acid β-glucosidase gene has been mapped to mouse chromosome 3 and its fine structure has been determined (*10*). Only the structural gene is present in the mouse genome. The exons of the human and murine genes are highly similar with 84% overall nucleotide identity in the protein coding, 54% in the 5' non-coding and 78% in the 3' non-coding regions. The positions of the 20

intron/exon junctions in the structural genes from each species were precisely conserved (10). Compared to the human sequence: 1) a single amino acid deletion, His[273], was predicted in the mouse sequence, 2) 64% of the nucleotide substitutions resulted from third position changes, 3) 100% of the cysteines, 91% of prolines and 89% of glycines were preserved, and 4) the greatest variation occurred in the N-terminal 60% of the predicted amino acid sequences.

Northern analyses of poly (A)[+] RNAs from Hela and cultured skin fibroblasts from normal individuals and Gaucher disease patients indicated three acid β-glucosidase specific mRNAs of ~5.6, 2.5 and 2.0 kb (7,11). The longest mRNA was thought to represent an unspliced or partially spliced nuclear transcript whereas the shorter mRNAs were suggested by S1 nuclease analyses to arise from alternate transcription initiation as well as alternative polyadenylation (8,11). Primer extension studies indicate that the major full length mRNA is about 2.5-2.6 kb and has 250 bp of 5' and 650 bp of 3' untranslated sequence including a 100 bp poly (A) tail (7,12).

An unusual feature of the human acid β-glucosidase cDNA was the presence of two 5' in-frame ATGs as possible translation initiation codons (12-15). These two ATGs begin 57 and 118 nucleotides 5' upstream of the first nucleotide of the codons for the mature polypeptide. The possible translation products encoded by sequences downstream from these ATGs are markedly different. If translation was initiated from the most 5' ATG, a very hydrophilic, charged sequence of 19 amino acids would be followed by a typical hydrophobic signal sequence of 20 amino acids which begins at the second ATG. This downstream ATG is preceded by a Kozak consensus sequence for translation initiation. Both ATGs were efficiently translated in a cell free system (8). Site-directed mutagenesis and expression studies of cDNAs containing either one or both ATGs demonstrated that functional acid β-glucosidase of the appropriate size was synthesized from these retroviral constructs which had been transfected into human Gaucher disease fibroblasts or NIH3T3 cells (16,17).

Acid β-Glucosidase Gene Mutations in Gaucher Disease. Interest in the genetic mutations causal to Gaucher disease derives from the need to explain the marked heterogeneity of phenotypes within and among the variants. The need for such genetic markers of disease severity (18,19) has had increasing practical (i.e., prognostic) and basic (structure/function) import. The numerous mutations identified to date are summarized in Figure 1 [reviewed in (3)]. Most of these are rare or have been found only in single families, but the N370S, 84GG (a G insertion at cDNA base 84), L444P and IVS2[+1] (a splice junction mutation) account for about 75, 15, 7 and 3% of Gaucher disease alleles, respectively (3).

Molecular Biology of Prosaposin. Studies of saposin B and saposin C showed that 1) each was derived by extensive post-translational proteolytic processing of a large molecular weight precursor (20,21), 2) the genes for both map to the same portion of chromosome 10 (22,23) and 3) both are encoded by the same gene in humans (24-27) and rats (28). The cDNA for "prosaposin" encodes four regions of high amino acid similarity (>80%) (Figure 2). These predicted saposins have been demonstrated by their partial or complete amino acid sequences

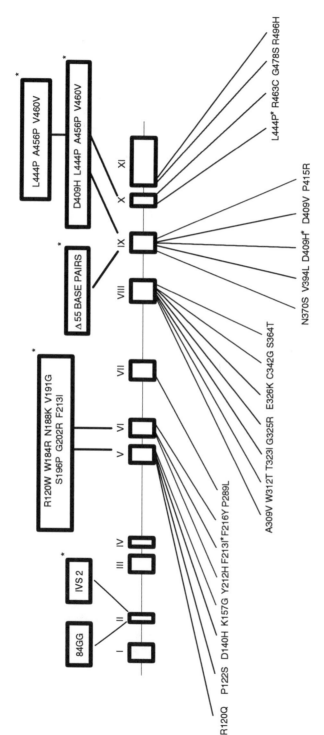

FIGURE 1: Mutations in the acidβ-glucosidase gene causing Gaucher disease. The approximate location of the mutations in the exons (roman numerals I-XI) and introns are shown. Alleles encoding missense mutations are shown below and more complex alleles are shown above the schematic genomic structure. The asterisks designate alleles which mimic pseudogene mutations. The mutations are numbered by the amino acid changed in the mature polypeptide (497 residues). 84GG, IVS 2 and Δ55 indicate a G insertion at nucleotide 84 of the cDNA, a splice junction G to A transition and a 55 bp deletion, respectively. The alleles N370S, L444P, 84GG, IVS 2 and R463C are the most common and have decreasing frequencies in the approximate order listed.

determined from isolated proteins (29-33). The only major difference in the amino acid sequences predicted by the full-length cDNAs (25,26) was the presence of an in-frame 9 bp insertion (3 amino acids) (26) from lung and skin fibroblast libraries. This may result from alternative mRNA splicing, but the additional amino acids were not present in the chemically determined sequence of saposin B (34). Similar to the acid β-glucosidase mRNA, two species of human prosaposin mRNA arise from alternative polyadenylation (27) and the levels of mRNA expression are concordant with acid β-glucosidase mRNA; i.e., the levels of prosaposin mRNA were higher in normal skin fibroblasts than in B-cells and the mRNA levels are higher in the corresponding cells from Gaucher disease patients. The organization (35,36) and sequence (36) of the ~20 kb prosaposin gene are shown schematically in Figure 2. Mutations in the saposin C region resulting in a Phe for Cys substitution and in the initiating codon for "prosaposin" have been reported in patients with the Gaucher-like disease (37-39).

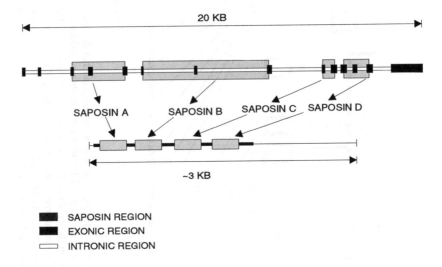

Figure 2: Schematic diagram of the prosaposin gene (top) and cDNA (bottom). The exons of the genomic sequence (~20 kb) are shown as black rectangles and introns are depicted as the open bars. The structure of the extreme 5' end of the gene including the first indicated exon has not been fully defined. The regions encoding the individual saposins include exonic and intronic sequences which are enclosed in larger shaded rectangles. The cDNA (~3 kb) is shown below in relation to the genomic structure. The mature saposins are derived by proteolytic processing of an intact prosaposin protein.

Characterization of the Glucosylceramide Cleavage Complex

Since the demonstration of *in vitro* glucosylceramide cleaving activity in mammalian tissues (*40,41*), it has become clear that the enzyme, acid β-glucosidase, and other protein components, activator(s) or saposins, are required for the *in vivo* degradation of this lipid (*42-46*). The current understanding of the interactions of these components of the glucosylceramide cleavage system is incomplete.

Acid β-Glucosidase. Acid β-glucosidase has been purified from a variety of species, but human tissues have been the primary source for many investigations (*47-53*). The enzyme is a homomeric that is glycoprotein encoded by its structural locus on chromosome 1 (*4,54,55*). Partial (*56*) and complete (*57*) amino acid sequences have been determined chemically for the placental enzyme and deduced from the nucleotide sequences of the cDNAs isolated from fibroblasts, hepatic and lymphoid libraries (*13-15,57*). Except for cloning artifacts, the predicted amino acid sequences were identical. The deduced amino acid sequence from the mouse cDNA indicated high (88%) identity to the human protein. These proteins have no apparent homology to yeast and plant β-glucosidases. The mature human polypeptide consists of 497 amino acids with a calculated molecular weight of 55,575. The pure glycosylated enzyme from placentae has a molecular weight of about 65,000. Aerts *et al.* (*58*) described a high molecular weight form of acid β-glucosidase in human cells and urine, termed glucocerebrosidase II, which was shown to be an enzyme/saposin C aggregate. The native state of acid β-glucosidase in cells and tissues has not been completely defined, due to its requirement for detergent or organic solvent extraction for solubilization. Molecular weights, estimated by sedimentation and molecular exclusion chromatography, have varied from about 60,000 to 450,000 (*47-53,56,59,60*). *In situ* radioinactivation estimates of molecular weight have been consistent with either monomeric or dimeric structures (*59-62*). The calculated pI value (7.2) of the polypeptide is consistent with that (pI=7.3-7.8) obtained with the pure placental enzyme containing only a neutral mannosyl oligosaccharide core (*63*). The protein has about 11% leucine residues and 45% non-polar amino acids but transmembrane domains are not present in the mature polypeptide. Proteolytic digestion studies of acid β-glucosidase obtained by *in vitro* translation in the presence of microsomal membranes showed the absence of a large cytoplasmic domain (*64*). These findings and those of Imai (*65*) indicate that acid β-glucosidase is a peripheral membrane protein. No obvious basis for its tight membrane association is evident from the primary sequence.

The human and murine sequences have conserved placement of the five N-glycosylation consensus sequences, but only those at amino acids 146-148 and 462-464 were identical in sequence (*10*). Takasaki *et al* (*66*) have shown typical bi- and tri-antennery complex N-linked oligosaccharides on the human placental acid β-glucosidase. However, the less common NeuAcα2→3Galβ1 linkages were present whereas NeuAcα2→6Galβ1 linkages were not (*66*). Endoglycosidase F digestion studies of human fibroblast acid β-glucosidase showed that only four of the potential glycosylation sites are occupied (*64*). Expression of the human cDNA in prokaryotes indicated that glycosylation is essential for the development of a catalytically active enzyme form (67), but it was not required to maintain activity in cells (68). Using site-directed mutagenesis, the first four glycosylation sites were shown to be normally occupied and that glycosylation of the first site was required for development of an active conformer (69).

Saposins C and A. Saposin C has been designated heat stable factor, co-β-glucosidase and SAP-2 (sphingolipid activator protein 2). The physiological importance of saposin C was proved by glucosylceramide storage and a Gaucher disease-like phenotype in patients with a deficiency of saposin C cross reacting immunological material and normal *in vitro* acid β-glucosidase activity (70,71). Recently, a second protein activator of *in vitro* acid β-glucosidase activity has been described (29). This protein has been designated saposin A. These saposins derive from a single gene product by extensive proteolytic and glycosidic post-translational processing.

Saposins A and C have been purified to homogeneity from normal and Gaucher disease spleens (33,34,44,46) and their amino acid sequences determined (24-27,29,46). These 80 amino acid (MW ~ 8950 Da) peptides are very acidic (pI ~ 4.2) glycoproteins (25). The single N-linked glycosylation consensus sequence in saposin C is occupied (25,72). Saposin A has two N-glycosylation consensus sequences and both are occupied (29). Very high levels of saposin C and A have been found in Gaucher disease tissues (73). The acid β-glucosidase activating function may be localized to residues 41-80 of the mature saposin C (74).

Post-Translational Processing of Acid β-Glucosidase. Biosynthesis of acid β-glucosidase has been investigated in cultured porcine kidney cells and human skin fibroblasts (64,75-78). The three forms of the enzyme present in extracts of these cells derive from a single polypeptide chain by differential post-translational oligosaccharide remodeling (78). By metabolic labeling the first detectable form of the acid β-glucosidase (64,75) contains high mannosyl chains (64,75). This initial form was transformed into a higher molecular weight species containing at least one complex chain over 1 to 24 hr (64,75) and the transformation was due to the remodeling of the oligosaccharide chains (77). The final lysosomal glycosylated form had $M_r = 59,000$. Signal sequence clipping is the only post-translational proteolytic processing (79). The half-life of acid β-glucosidase in cultured skin fibroblasts was estimated to be about 60 hours which is relatively short for lysosomal enzymes (75). Trafficking of acid β-glucosidase to the lysosome is independent of the mannose-6-phosphate receptor system.

Incomplete oligosaccharide processing of the mutant acid β-glucosidases has been observed in fibroblasts from patients with Gaucher disease variants (77,78,80-82). These abnormalities probably reflect the differential stabilities of the mutant proteins or their altered presentation to the processing enzymes. The presence of two different mutant acid β-glucosidase alleles in cells from many patients makes correlations of these altered properties to a genotype difficult in natural sources. Recreation of the individual mutations and expression of each in heterologous systems has begun to indicate regions on the protein which alter the stability of the mutant proteins (83,84) (Figure 3).

Biosynthesis and Processing of Saposin C and A. Fujibayashi and Wenger (20,21) used metabolic labeling to demonstrate extensive proteolytic and glycosidic

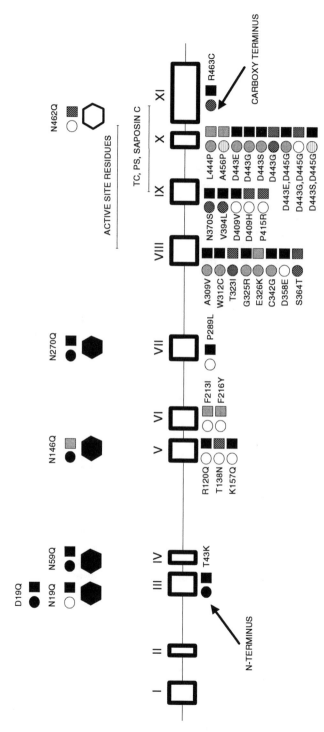

Figure 3: Functional map of acid β-glucosidase. The solid filled symbols indicate full or nearly full function for the mutant protein. Various partial fills represent more dysfunctional proteins. Circle and squares indicate catalytic activity and proteolytic/thermal stability, respectively. Filled or empty hexagons indicate normally occupied or unoccupied glycosylation consensus sites as determined by site-directed mutagenesis. The amino acid numbering corresponds to the residue in the mature enzyme protein sequence.

processing of saposin C in human skin fibroblasts. Very similar results were reported for the biosynthesis of saposin B (21). This would be expected since saposin C and saposin B are encoded by the same gene and mRNA (see below). However, the sequence of events involved in proteolytic processing of the "prosaposin" precursor and their potential tissue specificity have been defined. The highly homologous precursor produced in rat analogue (SGP-1) (28) is not proteolytically processed before its secretion into the culture media of Sertoli cells.

Enzymology of Acid β-Glucosidase

Kinetic studies of acid β-glucosidase have been complicated by its membrane association and its requirements for detergents, delipidation for solubilization and specific amphiphiles for reconstitution of enzymatic activity. The enzyme's natural substrates, glucosylceramides (85), are insoluble in aqueous media and require a lipoidal environment for dispersion.   Consequently, the assays for acid β-glucosidase are conducted in heterogeneous dispersions whose varying physical compositions and phase states influence the physical state and activity of the enzyme as well as its interactions with effector molecules. Because of the in vivo membrane association of acid β-glucosidase, investigators reasoned that specific membrane-derived lipids might be required for enzymatic activity. The acid β-glucosidase activity toward 4-methylumbelliferyl-β-$\underline{D}$-glucopyranoside (4MU-Glc) is enhanced by negatively charged phospholipids in the series, phosphatidylserine > phosphatidylinositol > phosphatidic acid, whereas phosphatidyl-ethanolamine or -choline does not affect enzymatic activity (44,56,86-90). Similar effects were found in detergent-free systems with glucosylceramide incorporated into unilamellar liposomes (91). Glew and coworkers delineated structure/activity relationships between   acid   β-glucosidase   and   negatively-charged   phospholipids   or glycosphingolipids (92-98). With a series of phosphatidylglycerols containing homogeneous fatty acid chains in the sn-1 and sn-2 positions, decreasing activation of acid β-glucosidase activity was obtained with increasing chain lengths from di-C_{12} to di-C_{18} saturated derivatives. Characterization of the lipids associated with acid β-glucosidase from different tissues could provide an interesting insight into the enzyme's in vivo functional hydrophobic environment.

The activating effects of saposin C on the hydrolysis of substrates by acid β-glucosidase have requirements for negatively charged phospholipids or glycosphingolipids (44,99).   These effects are thought to be mediated by conformational changes in acid β-glucosidase so that the interaction of phosphatidylserine conforms the enzyme into a "poised" form for interaction with saposin C (44). The use of inhibitory monoclonal antibodies has provided direct support for saposin C producing a conformational change in acid β-glucosidase that enhances catalytic activity (100). Similar kinetic results were reported for the effects of saposin A (29). It is not clear if saposin C and A bind to the same, overlapping (101) or different sites (100).

Properties of the Active Site. The natural substrates for acid β-glucosidase are mixtures of N-acylsphingosyl-1-O-β-$\underline{D}$-glucosides, glucosylceramides, with differing fatty acid acyl and sphingosyl moieties depending upon the tissue source (102,103). The synthetic substrate, 4-methylumbelliferyl-β-$\underline{D}$-glucopyranoside

(4MU-Glc), provides excellent estimates of acid β-glucosidase activity. The $K_m$app for 4MU-Glc is about 1.5-4 mM, about 100-fold greater than the $K_m$app for glucosylceramides. The enzyme activity is diprotic and has a pH optimum of about 5.5, that varies with composition of the assay mixture. The active site is specific for D-glucose since L-glucosylceramide is not hydrolyzed (*104*). D-glucosyl-D-*erythro*-ceramides are better substrates than the corresponding D-glucosyl-L-*threo* derivatives (*91*). Maximal *in vitro* hydrolytic rates are achieved with glucosylceramides containing fatty acids of 8 to 18 carbon atoms (*105*). The length of the fatty acid acyl chain had small effects on the $K_m$app values (*91,105*). The loss of the C4-C5 *trans* double bond in the sphingosyl moiety of the ceramide greatly reduces substrate affinity and hydrolysis (*91,105-107*). The β-glucoside derivatives of ceramides, alkyl-umbelliferyls, p-nitrophenyls and 2,3-di-alkyl-*sn*-glycerols (*105,108*) have k_{cat} values that differ only by 2-fold, but their respective apparent K_m values differ by 2 to 3 orders of magnitude. These findings indicate that the structure of the leaving group for 1-O-β-glucosides has little influence on inducing the conformational change needed for maximal catalytic power of the enzyme (*107*).

Based their results (*11,83,88,105,107,109,110*) and previous work (*111,112*) of substrate specificity and inhibitor analyses, Grabowski and coworkers proposed a kinetic model of the active site of acid β-glucosidase which includes three binding sites for the glycon head group, the sphingosyl moiety and the fatty acid acyl chain of glucosyl ceramide. Interaction at the glycon binding site requires a specific bonding with each hydroxyl group (*110*) on glucose while the sphingosyl and fatty acid acyl binding sites accommodate hydrophobic chains up to about 16 carbon atoms in length. These latter sites have been proposed to consist of multiple subsites that bind single methylene groups from the alkyl chains of substrates and inhibitors (*88,105,113*). This model is similar to those proposed for the binding of globotriaosylceramide to saposin B (*114*) and phospholipids to phospholipid transfer protein (*115*). Although supporting structural data will be required for these models, they may serve as prototypes for binding sites on lipid transfer proteins and partial or complete active sites of glycosphingolipid hydrolases.

The specificity of glycon binding was assessed by apparent binding constants for a series of epimers and substituted glucose derivatives (*110*). The enzyme binds β-glucose poorly, but has high affinity for 5-amino derivatives, i.e., the nojirimycins (*110*). Only the gluco derivatives of nojirimycin bound to the enzyme: i.e., the galactodeoxynojirimycin and deoxymannojirimycin even in millimolar concentrations did not inhibit the enzyme (*110*). These results and the lack of inhibition by galactosylsphingosine (*88*) showed the importance of the C2 and C4 hydroxyl configuration of the hexose. Alkyl-glycons were extremely powerful inhibitors of acid β-glucosidase (*88,105,110*), i.e., the N-alkyl-deoxynojirimycins and -β-glucosylamines had K_i values in the nanomolar and picomolar ranges, respectively (*105,107,116*). Since the additional free energies of binding were additive for the alkyl chain and the glycon head group, the existence of individual subsites for each methylene group up to about 16 was inferred to be within the active site (*105*). The N-alkyl-glucosylamines are likely reaction intermediate analogues and the transition state requires a conformational change of the enzyme during substrate hydrolysis (*107*). Beyond this, the reaction mechanism for acid β-glucosidase substrate hydrolysis is poorly defined.

Based on interactions with conduritol B epoxide derivatives and by analogy to other β-glucosidases, the catalytic mechanism may involve the sequential cleavage of the β-glucosidic linkage, release of ceramide, donation of water to a glucose-enzyme complex and the release of β-glucose (*117*). Recent use of site-directed mutagenesis and heterologous expression has provided some initial data to begin localization of portions of the enzyme which are responsible for catalytic activity (Figure 3).

Characterization of Mutant/Mutated Acid β-Glucosidases

Expression studies of alleles from Gaucher disease patients or mutated cDNAs are required to provide evidence for disease association. Such studies are also important for structure/function studies related to catalytic activity, substrate/inhibitor specificity, proteolytic stability and phospholipid or saposin C and A binding. Sufficient expression levels of the normal and mutagenized acid β-glucosidase cDNAs have been achieved using the baculovirus and other expression systems (*16,17,118-120*) to permit such comparative analyses. Glycosylated acid β-glucosidase with intact active site function (*118*) and normal signal peptide cleavage (*118,119*) was expressed from the normal cDNA that included only the more 3' (*118*) or both of the 5' ATGs (*121*). Importantly, cotranslational glycosylation in important to the development of catalytically active conformers of the enzyme (*122*). Stable enzyme protein lacking catalytic activity was expressed in bacteria or in eukaryotic systems in the presence of the glycosylation inhibitor, tunicamycin (*67*). In addition, site-directed mutagenesis studies showed that oligosaccharide occupancy or hydrophilicity of the first sequon was required for synthesis of a catalytically competent protein (*122*).

Grace *et al.* (*83,122,123*) have expressed several β-glucosidase cDNAs containing individual amino acid substitutions for several mutations found in Gaucher disease patients. Detailed analyses of inhibitor binding, phosphatidylserine activation and saposin C interaction with such mutant acid β-glucosidases provided an initial functional map of this enzyme (Figure 3). Although several naturally occurring mutations result in highly unfavorable amino acid replacements, some highly conservative or neutral substitutions also resulted in severe catalytic and/or stability defects. For example, the enzymes, D358E, S364T, N370S and V394L, all had defective catalytic activity. As referenced to the amount of cross-reacting immunological material, the latter three mutant enzymes had about 4-8 fold decreased catalytic rate constants whereas D358E was inactive. Conduritol B epoxide (CBE) and deoxynojirimycin (DNM) are covalent and reversible active site-directed inhibitors of acid β-glucosidase, respectively (*83*). As determined by IC50 values, CBE had decreased inhibitory potency with the N370S and V394L enzymes, but normal values with the S364T enzyme. Similarly, the IC50 values for DNM were increased for N370S, but were normal for S364T and V394L. These results suggest different residues within enzyme's active site are involved in the interaction with CBE and DNM. The more drastic mutant enzymes, L444P and R463C, interacted normally with these inhibitors. In comparison, the diminished responsiveness to phosphatidylserine and saposin C of the L444P and R463C, respectively, indicate important areas for interaction of these activators in the regions surrounding these residues (Figure 3).

Previously a Br-CBE binding site was identified as D443 (*124*). Substitution of glutamate, serine or glycine at this residue demonstrated that D443 is a binding site for Br-CBE but that D443 is not the nucleophile in catalysis. To exclude the possibility that D445 could function as an alternative nucleophile, glutamate, serine or glycine were substituted at this residue alone or in all combinations of these amino acids at residue 443. The finding of residual activity and Br-CBE inhibition of the resultant enzymes expressed using the baculovirus expression system excluded residue 445 as a potent alternative nucleophile. Misassignment of glycosidic nucleophiles has been previously observed with conduritol derivatives and β-galactosidase (*125,126*). Continued studies combining affinity labeling with transition state analogues and site-directed mutagenesis should continue to clarify the functional organization of human acid β-glucosidase.

ACKNOWLEDGEMENTS

The research described herein was supported by grants to GAG from the National Institutes of Health (DK 36729), the March of Dimes--National Birth Defects Foundation (1-857), the National Gaucher Foundation (NGF #19), the General Clinical Research Resources Branch of the National Institutes of Health (RR-71) and the Children's Hospital Research Foundation of Cincinnati, Ohio.

LITERATURE CITED

1. Gatt, S. **1966** *J.Biol.Chem.* 241, 3724-3730.
2. Yavin, Y.; Gatt, S. **1969** *Biochemistry* 8, 1692-1698.
3. Beutler, E.;Grabowski, G.A. *The Metabolic Basis of Inherited Disease*; Ed. 7th ;Eds.; Scriver, C.R.; Beaudet, A.L.; Sly, W.S.; Valle, D. McGraw-Hill: New York, 1993, in press.
4. Horowitz, M.; Wilder, S.; Horowitz, Z.; Reiner, O.; Gelbart, T.; Beutler, E. **1989** *Genomics* 4, 87-96.
5. Barneveld, R.A.; Keijzer, W.; Tegelaers, F.P.W.; Ginns, E.I.; Geurts van Kessel, A.; Brady, R.O.; Barranger, J.A.; Tager, J.M.; Galjaard, H.; Westerveld, A.; Reuser, A.J.J. **1983** *Hum.Genet.* 64, 227-231.
6. Devine, E.A.; Smith, M.; Arredondo-Vega, F.X.; Shafit-Zagardo, B.; Desnick, R.J. **1982** *Prog.Clin.Biol.Res.* 95, 511-534.
7. Reiner, O.; Horowitz, M. **1988** *Gene* 73, 469-478.
8. Reiner, O.; Wigderson, M.;Horowitz, M. *Lipid Storage Disorders. Biological and Medical Aspects*; Eds.; Salvayre, R.; Douste-Blazy, L.; Gatt, S. Plenum Press: New York, 1988, pp.29-39.
9. Dynan, W.S. **1986** *Trends Genet.* 2, 196-197.
10. O'Neill, R.R.; Tokoro, T.; Kozak, C.A.; Brady, R.O. **1989** *Proc.Natl.Acad.Sci.USA* 86, 5049-5053.

11. Graves, P.N.; Grabowski, G.A.; Ludman, M.D.; Palese, P.; Smith, F.I. **1986** *Am.J.Hum.Genet.* 39, 763-774.
12. Wigderson, M.; Firon, N.; Horowitz, Z.; Wilder, S.; Frishberg, Y.; Reiner, O.; Horowitz, M. **1989** *Am.J.Hum.Genet.* 44, 365-377.
13. Graves, P.N.; Grabowski, G.A.; Eisner, R.; Palese, P.; Smith, F.I. **1988** *DNA* 7, 521-528.
14. Sorge, J.; West, C.; Westwood, B.; Beutler, E. **1985** *Proc.Natl.Acad.Sci.USA* 82, 7289-7293.
15. Reiner, O.; Wilder, S.; Givol, D.; Horowitz, M. **1987** *DNA* 6, 101-108.
16. Sorge, J.; Kuhl, W.; West, C.; Beutler, E. **1987** *Proc.Natl.Acad.Sci.USA* 84, 906-909.
17. Choudary, P.V.; Barranger, J.A.; Tsuji, S.; Mayor, J.; LaMarca, M.E.; Cepko, C.L.; Mulligan, R.C.; Ginns, E.I. **1986** *Mol.Biol.Med.* 3, 293-299.
18. Theophilus, B.; Latham, T.; Grabowski, G.A.; Smith, F.I. **1989** *Am.J.Hum.Genet.* 45, 212-225.
19. Zimran, A.; Sorge, J.; Gross, E.; Kubitz, M.; West, C.; Beutler, E. **1989** *Lancet* 2, 349-352.
20. Fujibayashi, S.; Wenger, D.A. **1986** *J.Biol.Chem.* 261, 15339-15343.
21. Fujibayashi, S.; Wenger, D.A. **1986** *Biochim.Biophys.Acta* 875, 554-562.
22. Inui, K.; Kao, R.-T.; Fujibayashi, S.; Jones, C.; Morse, H.G.; Law, M.L.; Wenger, D.A. **1985** *Hum.Genet.* 69, 197-200.
23. Fujibayashi, S.; Kao, R.-T.; Jones, C.; Morse, H.; Law, M.; Wenger, D.A. **1985** *Am.J.Hum.Genet.* 37, 741-748.
24. O'Brien, J.S.; Kretz, K.A.; Dewji, N.; Wenger, D.A.; Esch, F.; Fluharty, A.L. **1988** *Science* 241, 1098-1101.
25. Gavrieli-Rorman, E.; Grabowski, G.A. **1989** *Genomics* 5, 486-492.
26. Nakano, T.; Sandhoff, K.; Stumper, J.; Christomanou, H.; Suzuki, K. **1989** *J.Biochem.* 105, 152-154.
27. Reiner, O.; Dagan, O.; Horowitz, M. **1989** *J.Mol.Neuroscience* 1, 225-231.
28. Collard, M.W.; Sylvester, S.R.; Tsuruta, J.K.; Griswold, M.D. **1988** *Biochemistry* 27, 4557-4564.
29. Morimoto, S.; Martin, B.M.; Yamamoto, Y.; Kretz, K.A.; O'Brien, J.S.; Kishimoto, Y. **1989** *Proc.Natl.Acad.Sci.USA* 86, 3389-3393.
30. Kleinschmidt, T.; Christomanou, H.; Braunitzer, G. **1987** *Biol.Chem.Hoppe Seyler* 368, 1571-1578.
31. Dewji, N.N.; Wenger, D.A.; O'Brien, J.S. **1987** *Proc.Natl.Acad.Sci.USA* 84, 8652-8656.
32. Morimoto, S.; Martin, B.M.; Kishimoto, Y.; O'Brien, J.S. **1988** *Biochem.Biophys.Res.Commun.* 156, 403-410.
33. Sano, A.; Radin, N.S.; Johnson, L.I.; Tarr, G.E. **1988** *J.Biol.Chem.* 263, 19597-19601.
34. Kleinschmidt, T.; Christomanou, H.; Braunitzer, G. **1988** *Biol.Chem.Hoppe Seyler* 369, 1361-1365.
35. Holtschmidt, H.; Sandhoff, K.; Furst, W.; Kwon, H-Y.; Schnabel, D.; Suzuki, K. **1991** *FEBS Lett.* 280, 267-270.
36. Gavrieli-Rorman, E.; Scheinker, V.; Grabowski, G.A. **1992** *Genomics* 13, 312-318.

37. Schnabel, D.; Schröder, M.; Sandhoff, K. **1991** *FEBS Lett.* 284, 57-59.
38. Maret, A.; Salvayre, R.; Troly, M.; Douste-Blazy, L. **1990** *Enzyme* 43, 99-106.
39. Carlson, D.E.; Busuttil, R.W.; Giudici, T.A.; Barranger, J.A. **1990** *Transplantation* 49, 1192-1194.
40. Brady, R.O.; Kanfer, J.N.; Shapiro, D. **1965** *Biochem.Biophys.Res.Commun.* 18, 221-225.
41. Patrick, A.D. **1965** *Biochem.J.* 97, 17C-18C.
42. Ho, M.W.; O'Brien, J.S. **1971** *Proc.Natl.Acad.Sci.USA* 68, 2810-2813.
43. Peters, S.P.; Coffee, C.J.; Glew, R.H.; Lee, R.E.; Wenger, D.A.; Li, S.C.; Li, Y.T. **1977** *Arch.Biochem.Biophys.* 183, 290-297.
44. Berent, S.L.; Radin, N.S. **1981** *Arch.Biochem.Biophys.* 208, 248-260.
45. Iyer, S.S.; Berent, S.L.; Radin, N.S. **1983** *Biochim.Biophys.Acta* 748, 1-7.
46. Peters, S.P.; Coyle, P.; Coffee, C.J.; Glew, R.H. **1977** *J.Biol.Chem.* 252, 563-573.
47. Pentchev, P.G.; Brady, R.O.; Hibbert, S.R.; Gal, A.E.; Shapiro, D. **1973** *J.Biol.Chem.* 248, 5256-5261.
48. Furbish, F.S.; Blair, H.E.; Shiloach, J.; Pentchev, P.G.; Brady, R.O. **1977** *Proc.Natl.Acad.Sci.USA* 74, 3560-3563.
49. Murray, G.J.; Youle, R.J.; Gandy, S.E.; Zirzow, G.C.; Barranger, J.A. **1985** *Anal.Biochem.* 147, 301-310.
50. Choy, F.Y. **1986** *Anal.Biochem.* 156, 515-520.
51. Strasberg, P.M.; Lowden, J.A.; Mahuran, D. **1982** *Can.J.Biochem.* 60, 1025-1031.
52. Grabowski, G.A.; Dagan, A. **1984** *Anal.Biochem.* 141, 267-279.
53. Aerts, J.M.; Donker-Koopman, W.E.; Murray, G.J.; Barranger, J.A.; Tager, J.M.; Schram, A.W. **1986** *Anal.Biochem.* 154, 655-663.
54. Ginns, E.I.; Choudary, P.V.; Tsuji, S.; Martin, B.; Stubblefield, B.; Sawyer, J.; Hozier, J.; Barranger, J.A. **1985** *Proc.Natl.Acad.Sci.USA* 82, 7101-7105.
55. Devine, E.A.; Smith, M.; Arrendondo-Vega, F.; Shafit-Zagardo, B.; Desnick, R.J. **1982** *Cytogenet.Cell Genet.* 33, 340-344.
56. Osiecki-Newman, K.M.; Fabbro, D.; Dinur, T.; Boas, S.; Gatt, S.; Legler, G.; Desnick, R.J.; Grabowski, G.A. **1986** *Enzyme* 35, 147-153.
57. Tsuji, S.; Choudary, P.V.; Martin, B.M.; Winfield, S.; Barranger, J.A.; Ginns, E.I. **1986** *J.Biol.Chem.* 261, 50-53.
58. Aerts, J.M.; Donker-Koopman, W.E.; van Laar, C.; Brul, S.; Murray, G.J.; Wenger, D.A.; Barranger, J.A.; Tager, J.M.; Schram, A.W. **1987** *Eur.J.Biochem.* 163, 583-589.
59. Maret, A.; Salvayre, R.; Negre, A.; Douste-Blazy, L. **1981** *Eur.J.Biochem.* 115, 455-461.
60. Maret, A.; Potier, M.; Salvayre, R.; Douste-Blazy, L. **1983** *FEBS Lett.* 160, 93-97.
61. Choy, F.Y.; Woo, M.; Potier, M. **1986** *Biochim.Biophys.Acta* 870, 76-81.
62. Dawson, G.; Ellory, J.C. **1985** *Biochem.J.* 226, 283-288.
63. Ginns, E.I.; Brady, R.O.; Stowens, D.W.; Furbish, F.S.; Barranger, J.A. **1980** *Biochem.Biophys.Res.Commun.* 97, 1103-1107.

64. Erickson, A.H.; Ginns, E.I.; Barranger, J.A. **1985** *J.Biol.Chem.* 260, 14319-14324.
65. Imai, K. **1985** *J.Biochem.(Tokyo)* 98, 1405-1416.
66. Takasaki, S.; Murray, G.J.; Furbish, F.S.; Brady, R.O.; Barranger, J.A.; Kobata, A. **1984** *J.Biol.Chem.* 259, 10112-10117.
67. Grace, M.E.; Grabowski, G.A. **1990** *Biochem.Biophys.Res.Commun.* 168, 771-777.
68. Van Weely, S.; Aerts, J.M.; Van Leeuwen, M.B.; Heikoop, J.C.; Donker-Koopman, W.E.; Barranger, J.A.; Tager, J.M.; Schram, A.W. **1990** *Eur.J.Biochem.* 191, 669-677.
69. Berg, A.; Grace, M.E.; Grabowski, G.A. **1992**, in review.
70. Christomanou, H.; Aignesberger, A.; Linke, R.P. **1986** *Biol.Chem.Hoppe Seyler* 367, 879-890.
71. Christomanou, H.; Chabás, A.; Pämpols, T.; Guardiola, A. **1989** *Klin.Wochenschr.* 67, 999-1003.
72. Sano, A.; Radin, N.S. **1988** *Biochim.Biophys.Acta* 154, 1197-1203.
73. Morimoto, S.; Yamamoto, Y.; O'Brien, J.S.; Kishimoto, Y. **1990** *Proc.Natl.Acad.Sci.USA* 87, 3493-3497.
74. Weiler, S.; Carson, W.D.; Ohashi, T.; Kishimoto, Y.; Morimoto, S.; O'Brien, J.S.; Aerts, J.M.; Tager, J.M.; Barranger, J.A.; Tomich, J.M. **1989** *Am.J.Hum.Genet.* 45, A12.
75. Jonsson, L.M.V.; Murray, G.J.; Sorrell, S.H.; Strijland, A.; Aerts, J.F.G.M.; Ginns, E.I.; Barranger, J.A.; Tager, J.M.; Schram, A.W. **1987** *Eur.J.Biochem.* 164, 171-179.
76. Beutler, E.; Kuhl, W. **1986** *Proc.Natl.Acad.Sci.USA* 83, 7472-7474.
77. Bergmann, J.E.; Grabowski, G.A. **1989** *Am.J.Hum.Genet.* 44, 741-750.
78. Fabbro, D.; Desnick, R.J.; Grabowski, G.A. **1987** *Am.J.Hum.Genet.* 40, 15-31.
79. Furbish, F.S.; Blair, H.E.; Shiloach, J.; Pentchev, P.G.; Brady, R.O. **1977** *Proc.Natl.Acad.Sci.USA* 74, 3560-3563.
80. Ginns, E.I.; Brady, R.O.; Pirruccello, S.; Moore, C.; Sorrell, S.; Furbish, F.S.; Murray, G.J.; Tager, J.; Barranger, J.A. **1982** *Proc.Natl.Acad.Sci.USA* 79, 5607-5610.
81. Jonsson, L.M.; Murray, G.J.; Sorrell, S.H.; Strijland, A.; Aerts, J.F.; Ginns, E.I.; Barranger, J.A.; Tager, J.M.; Schram, A.W. **1987** *Eur.J.Biochem.* 164, 171-179.
82. Tager, J.M.; Aerts, J.M.; Jonsson, M.V.; Murray, G.J.; Van Weely, S.; Strijland, A.; Ginns, E.I.; Reuser, J.J.; Schram, A.W.;Barranger, J.A. *Enzymes of Lipid Metabolism II*; Eds.; Freysz, L.; Dreyfus, H.; Massarelli, R.; Gatt, S. Plenum Press: New York, 1986, pp.735-745.
83. Grace, M.E.; Graves, P.N.; Smith, F.I.; Grabowski, G.A. **1990** *J.Biol.Chem.* 265, 6827-6835.
84. Grace, M.E.; Newman, K.M.; Scheinker, V.; He, G-S.; Berg, A.; Grabowski, G.A. **1992**,in review.
85. Nilsson, O.; Svennerholm, L. **1982** *J.Neurochem.* 39, 709-718.
86. Mueller, O.T.; Rosenberg, A. **1979** *J.Biol.Chem.* 254, 3521-3525.

87.   Legler, G.; Liedtke, H. **1985** *Biol.Chem.Hoppe Seyler* 366, 1113-1122.
88.   Grabowski, G.A.; Gatt, S.; Kruse, J.; Desnick, R.J. **1984** *Arch.Biochem.Biophys.* 231, 144-157.
89.   Glew, R.H.; Daniels, L.B.; Clark, L.S.; Hoyer, S.W. **1982** *J.Neuropathol.Exp.Neurol.* 41, 630-641.
90.   Dale, G.L.; Villacorte, D.G.; Beutler, E. **1976** *Biochem.Biophys.Res.Commun.* 71, 1048-1053.
91.   Sarmientos, F.; Schwarzmann, G.; Sandhoff, K. **1986** *Eur.J.Biochem.* 160, 527-535.
92.   Basu, A.; Glew, R.H. **1984** *Biochem.J.* 224, 515-524.
93.   Basu, A.; Prence, E.; Garrett, K.; Glew, R.H.; Ellingson, J.S. **1985** *Arch.Biochem.Biophys.* 243, 28-34.
94.   Prence, E.M.; Garrett, K.O.; Glew, R.H. **1986** *Biochem.J.* 237, 655-662.
95.   Basu, A.; Glew, R.H.; Wherrett, J.R.; Huterer, S. **1986** *Arch.Biochem.Biophys.* 245, 464-469.
96.   Basu, A.; Glew, R.H. **1985** *J.Biol.Chem.* 260, 13067-13073.
97.   Gonzales, M.L.; Basu, A.; de Haas, G.H.; Dijkman, R.; van Oort, M.G.; Okolo, A.A.; Glew, R.H. **1988** *Arch.Biochem.Biophys.* 262, 345-353.
98.   Basu, A.; Glew, R.H.; Daniels, L.B.; Clark, L.S. **1984** *J.Biol.Chem.* 259, 1714-1719.
99.   Prence, E.; Chakravorti, S.; Basu, A.; Clark, L.S.; Glew, R.H.; Chambers, J.A. **1985** *Arch.Biochem.Biophys.* 236, 98-109.
100. Fabbro, D.; Grabowski, G.A. **1991** *J.Biol.Chem.* 266, 15021-15027.
101. Morimoto, S.; Kishimoto, Y.; Tomich, J.; Weiler, S.; Ohashi, T.; Kretz, K.A.; O'Brien, J.S. **1990** *J.Biol.Chem.* 265, 1933-1937.
102. Mansson, J.E.; Vanier, T.; Svennerholm, L. **1978** *J.Neurochem.* 30, 273-275.
103. Nilsson, O.; Svennerholm, L. **1982** *J.Lipid Res.* 23, 327-334.
104. Gal, A.E.; Pentchev, P.G.; Massey, J.M.; Brady, R.O. **1979** *Proc.Natl.Acad.Sci.USA* 76, 3083-3086.
105. Osiecki-Newman, K.; Fabbro, D.; Legler, G.; Desnick, R.J.; Grabowski, G.A. **1987** *Biochim.Biophys.Acta* 915, 87-100.
106. Vaccaro, A.M.; Kobayashi, T.; Suzuki, K. **1982** *Clin.Chim.Acta* 118, 1-7.
107. Greenberg, P.; Merrill, A.H.; Liotta, D.C.; Grabowski, G.A. **1990** *Biochim.Biophys.Acta* 1039, 12-20.
108. Glew, R.H.; Gopalan, V.; Hubbell, C.A.; Devraj, R.V.; Lawson, R.A.; Diven, W.F.; Mannock, D.A. **1991** *Biochem.J.* 274, 557-563.
109. Gatt, S.; Dinur, T.; Osiecki, K.; Desnick, R.J.; Grabowski, G.A. **1985** *Enzyme* 33, 109-119.
110. Osiecki-Newman, K.; Legler, G.; Grace, M.; Dinur, T.; Gatt, S.; Desnick, R.J.; Grabowski, G.A. **1988** *Enzyme* 40, 173-188.
111. Erickson, J.S.; Radin, N.S. **1973** *J.Lipid Res.* 14, 133-137.
112. Hyun, J.C.; Misra, R.S.; Greenblatt, D.; Radin, N.S. **1975** *Arch.Biochem.Biophys.* 166, 382-389.
113. Grabowski, G.A.; Osiecki-Newman, K.; Dinur, T.; Fabbro, D.; Legler, G.; Gatt, S.; Desnick, R.J. **1986** *J.Biol.Chem.* 261, 8263-8269.
114. Wynn, C.H. **1986** *Biochem.J.* 240, 921-924.

115. Van Paridon, P.A.; Visser, A.J.; Wirtz, K.W. **1987** *Biochim.Biophys.Acta* 898, 172-180.
116. Goldblatt, J.; Sacks, S.; Dall, D.; Beighton, P. **1988** *Clin.Orthop.* 94-98.
117. Grabowski, G.A.; Gatt, S.; Horowitz, M. **1990** *Crit.Rev.Biochem.Mol.Biol.* 25, 385-414.
118. Grabowski, G.A.; White, W.R.; Grace, M.E. **1989** *Enzyme* 41, 131-142.
119. Martin, B.M.; Tsuji, S.; LaMarca, M.E.; Maysak, K.; Eliason, W.; Ginns, E.I. **1988** *DNA* 7, 99-106.
120. Ohashi, T.; Chang, M.H.; Weiler, S.; Tomich, J.M.; Aerts, J.M.; Tager, J.M.; Barranger, J.A. **1991** *J.Biol.Chem.* 266, 3661-3667.
121. Sorge, J.A.; West, C.; Kuhl, W.; Treger, L.; Beutler, E. **1987** *Am.J.Hum.Genet.* 41, 1016-1024.
122. Grace, M.E.; Berg, A.; He, G-S.; Grabowski, G.A. **1992** *Pediatr.Res.* 31:133A.
123. Grace, M.E.; Berg, A.; He, G.S.; Goldberg, L.; Horowitz, M.; Grabowski, G.A. **1991** *Am.J.Hum.Genet.* 49, 646-655.
124. Dinur, T.; Osiecki, K.M.; Legler, G.; Gatt, S.; Desnick, R.J.; Grabowski, G.A. **1986** *Proc.Natl.Acad.Sci.USA* 83, 1660-1664.
125. Gebler, J.C.; Aebersold, R.; Withers, S.G. **1992** *J.Biol.Chem.* 267, 11126-11130.
126. Trimbur, D.E.; Warren, R.A.; Withers, S.G. **1992** *J.Biol.Chem.* 267, 10248-10251.

RECEIVED January 12, 1993

Chapter 7

The Mammalian Cytosolic Broad-Specificity β-Glucosidase

Robert H. Glew[1], Venkatakrishnan Gopalan[1,3], George W. Forsyth[2], and Dorothy J. VanderJagt[1]

[1]Department of Biochemistry, School of Medicine, University of New Mexico, Albuquerque, NM 87131–5221
[2]Veterinary Physiological Sciences, University of Saskatchewan, Saskatoon, Saskatchewan S7N 0W0, Canada

Mammals contain two β-glucosidases: lysosomal glucocerebrosidase and a cytosolic β-glucosidase which has a very broad-specificity and is most active at neutral pH. The physiologic function of the cytosolic β-glucosidase is obscure, but it does hydrolyze toxic plant glycosides found in the diet of man. We summarize insights which pertain to the enzyme's structure, properties, and function and discuss the issues of tissue distribution, isolation schemes, and catalytic properties, including transglucosylation. Finally, we suggest avenues of research to be taken which are likely to lead to a better understanding of the function and pathophysiologic significance of the cytosolic β-glucosidase.

Glycosidases are widespread in nature and are responsible for the catabolism of a variety of carbohydrate-containing compounds. These hydrolases catalyze chemical transformations at the C-1 position of carbohydrates, and a number of metabolic processes depend on these enzymes for their efficiency, selectivity and regulation. Alpha- and β-glycosidases catalyze the hydrolysis of α- and β-glycosidic bonds that occur in polysaccharides, glycoproteins, and glycolipids (1).

In some mammalian tissues there is a bimodal distribution of β-glucosidase activity between the lysosomal and cytosolic compartments. In the earliest studies β-glucosidase activity was found in both the lysosome-rich and the unsedimentable (cytosolic) fractions of rat kidney (2, 3). The lysosomal β-glucosidase, glucocerebrosidase, has a pH optimum in the acidic range, whereas the cytosolic β-glucosidase is most active in the pH range 6.0 - 7.0. It is now well established that the lysosomal β-glucosidase is the enzyme responsible for the hydrolysis of the glycosphingolipid, glucocerebroside, to glucose and ceramide (4). The

[3]Current address: Department of Biology, Yale University, 844 Kline Biology Tower, New Haven, CT 06511

deficiency of glucocerebrosidase is the biochemical basis for the sphingolipidosis called Gaucher disease (5, 6). Comprehensive information on a range of topics from purification of glucocerebrosidase to cloning of the gene is the subject of recent reviews (4 ,7). The other glucohydrolase is a cytosolic β-glucosidase with broad substrate specificity. In in vitro assays it catalyzes the hydrolysis of β-D-glucosides, β-D-galactosides, α-L-arabinosides and β-D-xylosides conjugated to the aglycones p-nitrophenol or 4-methylumbelliferone (8, 9). The function and pathophysiological significance of this neutral pH optimum glycohydrolase remain uncertain.

In animal models of renal disease, the levels of urinary cytosolic β-glucosidase were found to be elevated long before proteinuria was apparent (10-12). Soon thereafter, it was appreciated that the increased excretion of this enzyme in urine could be used as a diagnostic marker for the detection of renal damage and to forewarn of renal-transplant rejection (13). This finding provided the initial impetus to charaterize the cytosolic β-glucosidase. Also, since the various clinical phenotypes observed in patients with Gaucher disease could not be explained on the basis of a deficiency in glucocerebrosidase activity alone, many investigators have compared the two mammalian β-glucosidases in an effort to reveal possible evolutionary or functional similarities and also to determine if the cytosolic enzyme has any role in the etiology of Gaucher disease. Until recently, research on the cytosolic enzyme had been limited primarily to its purification from various sources and characterization of its kinetic properties.

The use of amphipathic compounds to probe the physical and chemical nature of hydrophobic effector sites on the guinea pig liver cytosolic β-glucosidase has furnished clues about possible physiological substrates and details of the reaction mechanism of the cytosolic enzyme (14). In addition, studies on the hydrolysis of aryl β-D-glucosides by the cytosolic β-glucosidase have provided insights regarding the rate-limiting step and the nature of the intermediates generated during catalysis. Finally, recent investigations have revealed that some naturally occurring toxic β-D-glucosides in the human food chain are substrates of the cytosolic β-glucosidase (15). In this review we attempt to condense the current fund of knowledge on the mammalian cytosolic β-glucosidase and also deliberate on new ideas and approaches to address the evasive question of the metabolic and pathophysiologic role of this enigmatic β-glucosidase.

The Cytosolic Broad Specificity β-Glucosidase

Standard β-Glucosidase Assays. Two particular nonphysiologic substrates have been widely utilized to measure β-glucosidase activity (Figure 1). The first is 4-methylumbelliferyl-β-D-glucoside (4-MUGlc) which when cleaved yields the aglycone 4-methylumbelliferone (4-MU) that is highly fluorescent at alkaline pH (16, 17). Alternatively, p-nitrophenyl-β-D-glucoside has been used as a β-glucosidase substrate in a spectrophotometric assay in which the release of the p-nitrophenol is estimated based on the absorbance of the phenolate ion at 400 nm (18). One can also measure the rate of glucose release from the glucoside substrates in a coupled assay system using hexokinase and glucose 6-phosphate dehydrogenase (19).

A continuous spectrophotometric assay for the β-glucosidase and one that opens the possibility for measuring pre-steady state kinetics has been reported recently (*20*). Lai and associates demonstrated that the neutral pH optimum cytosolic β-glucosidase catalyzes the hydrolysis of the plant glucosides L-picein (p-hydroxy-acetophenone-β-D-glucoside) and prunasin (D-mandelonitrile-β-D-glucoside) (Figure 2). The marked differences in the spectra of the substrate/product pairs of L-picein/p-hydroxyacetophenone and prunasin/mandelonitrile permit continuous monitoring of the β-glucosidase-catalyzed release of p-hydroxyacetophenone from L-picein and mandelonitrile from prunasin. K_m and V_{max} values obtained from the continuous spectrophotometric assays agreed well with values reported previously based on discontinuous assay procedures.

Tissue Distribution of Mammalian Cytosolic β-Glucosidase. There are wide differences between species and tissues in the distribution and content of the cytosolic β-glucosidase. In general, the richest source of the β-glucosidase in most vertebrates appears to be the liver. On the basis of assays performed using p-nitrophenyl-β-D-galactoside as the substrate, Distler and Jourdian (*21*) found the following specific activities of the broad-specificity β-glucosidase in the livers of various species: bovine - 39.0; porcine - 24.2; human - 15.3; guinea pig - 6.5; rabbit - 6.3; rat - 0.88; beaver - 0.83: and mouse - 0.20 nmole/min/mg. They also reported that the livers of birds, reptiles, and fish were completely devoid of this activity. In the cow, the broad-specificity β-glucosidase is abundant in liver, kidney, and intestinal mucosa, but absent from spleen, skeletal muscle, testes, and serum. In rabbits, the enzyme is predominantly distributed in the cytosol of liver and kidney (*22, 23*). In the guinea pig, using Western blot analysis, we found that the cytosolic β-glucosidase is present in the largest amounts in the liver, intestine, stomach, and spleen (Table I) (S. Macko and R. H. Glew, unpublished observation). Surprisingly, there is very little cytosolic β-glucosidase activity in mouse and rat liver; however, rat kidney is a rich source of the enzyme (*24*). This specific distribution pattern may relate to a metabolic role of the cytosolic β-glucosidase.

Isolation and Purification. The cytosolic β-glucosidase has been purified to homogeneity from the livers of man (*8*), calf (*25*), and guinea pig (*19*) and also from porcine kidney (*26*). In general, the purification schemes have taken advantage of the acidic and hydrophobic properties of the enzyme. Purification procedures developed in three different laboratories are reviewed in this section.
    The current protocol used in our laboratory for isolating guinea pig liver cytosolic β-glucosidase is described below (*27*). Liver tissue is minced and homogenized in 10 mM sodium phosphate buffer (pH 6.0) supplemented with the protease inhibitors aprotinin, leupeptin, pepstatin, and phenylmethylsulfonyl fluoride. The 100,000 x g (1h) supernatant is subjected to DEAE-cellulose and hydroxyapatite chromatography. The enzyme preparation is then applied to an octyl-Sepharose column that has been pre-equilibrated with 0.5 M $(NH_4)_2SO_4$ to enhance the interaction between the cytosolic β-glucosidase and the hydrophobic affinity resin. The enzyme can be eluted from the affinity column using 60% (v/v)

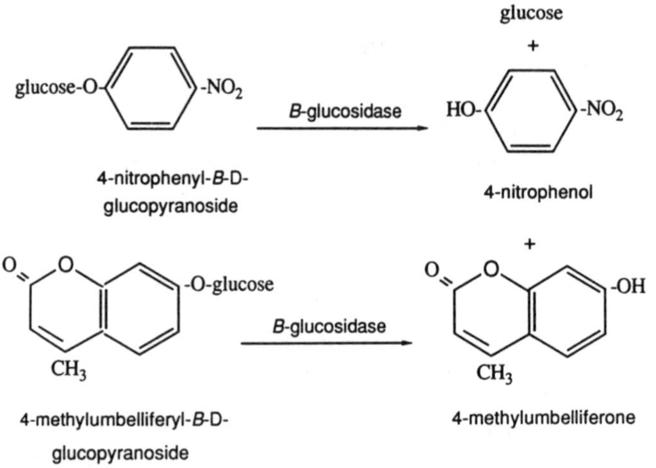

Figure 1. Two aryl-β-D-glucoside substrates used to measure β-glucosidase activity.

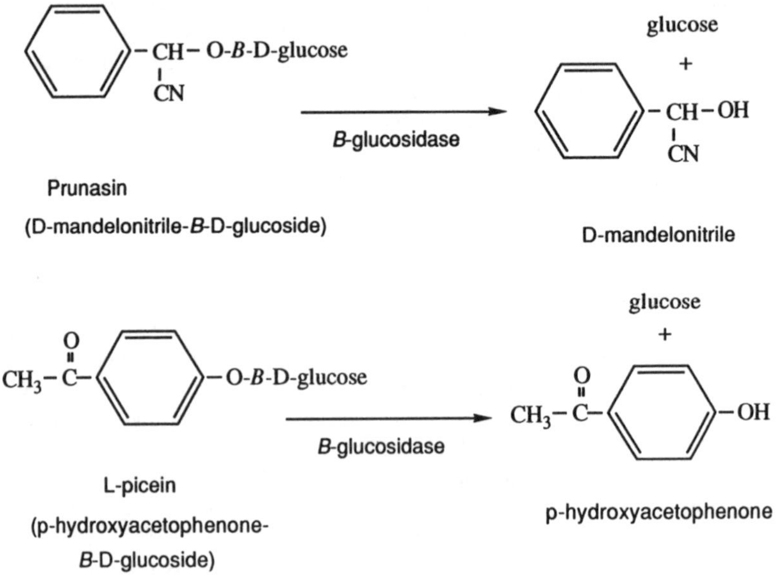

Figure 2. Two naturally occurring aryl β-D-glucosides.

ethylene glycol, 5% (v/v) n-butanol, or 2% (v/v) n-pentanol. The hydrophobic affinity column consistently yields the maximum fold purification. Gel filtration chromatography (Waters ProteinPak S300 column) can be used to remove the last traces of contaminating proteins. This treatment yields a homogeneous preparation when analyzed by SDS polyacrylamide gel electrophoresis. This purification scheme gives a 30% recovery and an overall 2000-fold purification. The isolated β-glucosidase retains full activity for several weeks when stored at -20°C in 60% (v/v) ethylene glycol.

Legler and Bieberich reported a rapid and efficient procedure for the isolation of the cytosolic β-glucosidase from calf liver (*25*). They subjected the high-speed supernatant to consecutive steps of acid precipitation (pH 5.0), heat treatment (53°C, 30 min), chromatography on DEAE-cellulose, and affinity

Table I. Content of Cytosolic β-Glucosidase Detected by Specific Antibody in Selected Guinea Pig Tissues

Tissue	β-Glucosidase Specific Activity (units/mg protein)	β-Glucosidase Antigen Level
Liver	48,900	+ + + + +
Small intestine	16,600	+ + + + +
Large intestine	5260	+ + +
Heart	1370	-
Lung	1050	-
Kidney	356	-
Stomach	18,900	+ + + +
Spleen	6360	+ + +
Brain	109	-

Guinea pig organ homogenates were centrifuged at 100,000 x g for 1 h, and the supernatants were chromatographed on octyl-Sepharose (23). The eluents were analysed for β-glucosidase activity using the standard MUGlc assay. One unit of activity corresponds to the hydrolysis of 1 nmol of MUGlc per h. For Western blot analysis, 4 μg of protein eluted from an octyl Sepharose column was separated by SDS-polyacrylamide gel electrophoresis, blotted to nitrocellulose, and exposed to rabbit IgG prepared against isolated guinea pig liver β-glucosidase. Antigen amounts were ranked on a scale of + + + + + as largest, and - as no antigen detected.

chromatography using a N-(9-carboxynonyl) deoxynojimycin-AH-Sepharose column.

Pocsi and Kiss isolated the broad-specificity β-glucosidase from an acetone powder of pig kidney (26). Following acid precipitation, electrofocusing was performed using a pH gradient from 4.3 to 7.5. Finally, molecular sieve chromatography on Sephacryl S-200 yielded pure cytosolic β-glucosidase.

The specific activity of the purified enzyme from guinea pig liver and calf liver is approximately 8 μmol/min/mg protein when assayed using 4-MUGlc. The broad-specificity nature of the enzyme has been indicated by coelution of the four glycosidic activities (i.e., β-D-glucosidase, β-D-galactosidase, β-D-xylosidase and α-L-arabinosidase) during the various chromatographic steps (19).

Physicochemical Properties

Size. Most mammalian β-glucosidases purified to date using buffers that have contained protease inhibitors have exhibited molecular masses in the 55 - 65 kDa range. These estimates have been derived from SDS-PAGE analysis and gel filtration chromatography under non-denaturing conditions. The similarity of the size estimates from the two procedures suggests that the enzyme functions in the monomeric state. Daniels and associates reported a sedimentation coefficient ($S_{w,20}$) of 4.44 $\pm$ 0.03 for the purified human liver cytosolic β-glucosidase (8). A Stokes radius of 31.0 Å was determined for the same human enzyme using the molecular sieve chromatographic method of Ackers (28).

Charge. The relatively high affinity of the cytosolic β-glucosidase for the anion exchanger DEAE-cellulose indicates that the protein possesses an overall anionic character at pH 6.0. In fact, most mammalian cytosolic β-glucosidases are relatively acidic proteins with pI values in the range 4.5 - 5.2 (16, 26).

Carbohydrate Content. The guinea pig liver β-glucosidase does not react in the periodate-Schiff base test and fails to bind to concanavalin A-Sepharose (16), suggesting that it is not a glycoprotein. However, a rigorous systematic study exploring the possibility of other post-translational modifications has not been reported.

Hydrophobicity. One of the earliest indications of the marked hydrophobic nature of the soluble β-glucosidase came from the observation that the liver enzyme has a high affinity for alkyl β-D-thioxylosides (29). These amphipathic glycosides were demonstrated to be potent reversible inhibitors of the human liver cytosolic β-glucosidase. This property has been utilized in the design of purification schemes that have included affinity chromatography on phenyl- or octyl-Sepharose or alkyldeoxynojirimycin-Sepharose (19, 25). Recent results from our laboratory support the existence of hydrophobic binding sites in the interior of the enzyme, specifically in proximity to, or part of, the active site (41).

Immunochemical Properties. It is widely accepted that isoenzymes generated from distinct gene loci exhibit differences in antigenicity (30). The failure of antibodies prepared against the placental lysosomal glucocerebrosidase to cross-react with human liver cytosolic β-glucosidase led Daniels and associates to conclude that the two enzymes are unrelated (*9*). Furthermore, rabbit polyclonal antibodies directed against the guinea pig liver cytosolic β-glucosidase did not recognize human spleen glucocerebrosidase (V. Gopalan and R. H. Glew, unpublished observation). However, this does not preclude the existence of specific amino acid sequences that are common to the two β-glucosidases.

Mechanism of Action of Mammalian Cytosolic β-Glucosidase

Sufficient kinetic data has been obtained by several investigators to formulate a basic mechanism for the hydrolysis of aryl glycosides by the cytosolic β-glucosidase. The mechanism of hydrolysis of β-D-glucosides by the mammalian β-glucosidases (both lysosomal and cytosolic) can be classified as a nucleophilic substitution with retention of anomeric configuration and is analogous to that of lysozyme and other glycosidases (*1, 31, 32, 33*). According to the proposed reaction mechanism, a proton donor (an AH group on the enzyme) first protonates the anomeric oxygen in the glycoside and facilitates the release of the aglycone moiety (Figure 3). Concomitantly, an oxocarbonium ion is generated which is stabilized by a negatively-charged residue in the active site of the enzyme (an Asp-β-COO⁻ or Glu-γ-COO⁻) until it is neutralized by a nucleophile. The C(1) atom of the glucosyl oxocarbonium ion exists in the sp^2 state, and it has been proposed that the carbocation could exist as a puckered ring or in a half-chair form (*1, 32*). The glucosyl oxocarbonium ion intermediate then interacts with the enzyme in a manner that permits stereospecific solvolysis to occur such that final attack by the solvent (H_2O or ROH) is from one side only, thereby ensuring the formation of the β-D-glucose product. Such stereospecific nucleophilic attack of the glucosyl carbocation is termed hydrolysis (if the acceptor is water) or transglucosylation (if the acceptor is any other hydroxylic compound such as n-butanol or benzyl alcohol).

While the mechanism proposed above involves the generation of a non-covalent intermediate during catalysis, the possibility of a covalent glucosyl-enzyme intermediate in the enzyme-catalyzed reaction cannot be ruled out. If such an intermediate were formed subsequent to the rate-limiting step it cannot be identified since it will not accumulate in appreciable amounts. Capon (*32*)has also argued that the rigid geometry of the protein may not permit formation of a covalent bond between the substrate and the enzyme. However, the situation could arise wherein the energy to be gained from forming the glycosyl-enzyme intermediate would serve to distort the protein in order to permit formation of a covalent linkage.

Two recent studies support the hypothesis that a partial covalent intermediate is actually formed during lysozyme catalysis. Hardy and Poteete reported that with bacteriophage T4 lysozyme, replacement of the critical Asp[20] by Cys produced an enzyme with near wild type specific activity (*34*). Muraki et al. showed that replacement of Asp[53] by Glu in human lysozyme reduced the activity to a few percent of that of the wild type (*35*). Taken together, these results suggest that the aspartate residue has a significant nucleophilic function that is more important than serving a mere electrostatic role. The general scheme proposed above for the enzymatic catalysis of glucoside hydrolysis may apply also to the mammalian cytosolic β-glucosidase, since it accommodates most of the experimental observations reported on this glycosidase.

pH-Dependence of the Enzyme-Catalyzed Reaction. The study of pH dependence of the activity of the calf liver cytosolic β-glucosidase (i.e., ln V_{max}/K_m versus pH) by Legler and Bieberich revealed the participation of two functional groups with pK_a values of 3.9 and 7.3 in the rate-limiting step (25). A carboxylate group with $pK_a = 3.9$ could provide the general base catalyst needed to stabilize the oxocarbonium ion intermediate, while a protonated acidic group with $pK_a = 7.3$ could protonate the anomeric oxygen of the glycoside substrate. Deprotonation of this acidic group ($pK_a = 7.3$) seems to be detrimental to the binding of the substrate to the enzyme as is reflected in the steep increase in K_m at pH values greater than 7.0. Legler and Bieberich also concluded from an analysis of ln V_{max} versus pH plots that ionization of the ES complex shifts the pK_a of the acidic group from 7.3 to 8.9 (25). They postulated that the latter effect was caused by strong shielding of this group from the aqueous environment of the assay medium.

Glew and associates documented that the activity of rat kidney β-glucosidase is inhibited almost completely when the sulfhydryl reagents iodoacetic acid, p-chloromercuribenzoate, N-ethylmaleimide or 5,5'-dithio-bis-(2-nitrobenzoic acid) are included in the standard assay medium (24). This suggests that the enzyme contains at least one functionally essential and accessible sulfhydryl group, the role of which remains unclear.

Evidence for a Double-Displacement Mechanism. The stereochemical outcome of the reaction is an important question in considering the mechanism of hydrolysis of any glycoside. The anomeric configuration of the glucose released on hydrolysis of pNP-β-D-glucoside was determined as follows (36). The hydrolysis of p-nitrophenyl-β-D-glucoside was monitored by measuring the rate of appearance of p-nitrophenol. Polarimetric measurement of D-glucose was used to verify the expected equimolar release of p-nitrophenol and β-D-glucose in the same β-glucosidase assay. The results confirmed that cleavage of the β-glucoside occurred with retention of anomeric configuration. Using gas-liquid chromatography we recently obtained evidence for the preservation of the anomeric configuration of the original β-D-glucoside molecule in the alkyl-β-D-glucoside product generated by the cytosolic β-glucosidase during alcoholysis (14). These observations point to catalysis of aryl-β-D-glucoside hydrolysis through a double-displacement · reaction mechanism involving two sequential S_n2 substitutions, and sequential liberation of products (aglycone release, followed by the discharge of the glycone).

Inhibitors. 1,5-Gluconolactone was one of the first potent glycosidase inhibitors to be reported (37). Daniels and associates showed that the human liver cytosolic β-glucosidase is inhibited by micromolar concentrations of 1,5-gluconolactone (8). The half-chair conformation of 1,5-gluconolactone has been confirmed by X-ray analysis (38). This observation supports the hypothesis that the inhibitory effects of 1,5-gluconolactone are attributable to structural similarities between it and the transition state (i.e., a half-chair glycosyl oxocarbonium ion) in the enzyme-catalyzed hydrolysis of glucosides (*1*). This result also supports the postulate that distortion of the substrate from a normal chair conformation to a half-chair structure favors the catalytic reaction.

Bromo conduritol B epoxide (1,2-anhydro-6-bromo-6-deoxy-myo-inositol, BrCBE) is a potent covalent irreversible inhibitor of the cytosolic β-glucosidase (*36*). Release of the bound inhibitor as (1,3,4/2,5,6)-6-bromocyclohexanepentol after denaturation of the enzyme and treatment with ammonia led Legler and Bieberich (*36*) to conclude that an oxygen ester linkage was probably formed between the epoxide inhibitor and a carboxylate residue in the active site of the enzyme (Figure 4). The essential general base catalyst involved in the enzymatic reaction could be provided by either a glutamate or aspartate residue. Isolation and sequencing of the peptide containing the active site is needed to confirm these speculations concerning the nature of the amino acids that are critical to the catalytic mechanism used by the cytosolic β-glucosidase.

Enzyme kinetic studies performed using inhibitor pairs comprised of compounds with charged groups and their neutral analogs have provided valuable information about the ionic character of the active site of the enzymes (*39*). Several observations pertaining to the interaction of inhibitors with the active site of the cytosolic β-glucosidase are worth mentioning in this regard. A comparison of the inhibitory potential of C_n-1-O-β-D-glucosides (i.e., alkyl-β-D-glucosides) versus the C_n-1-N-β-D-glucosides (i.e., alkyl β-D-glucosylamines) revealed that replacement of the anomeric oxygen with a nitrogen resulted in a thousand-fold increase in inhibitory potential (*25*). Furthermore, the K_I value for the cationic β-D-galactosyl pyridinium ion was 10 mM compared to the K_I value of 0.035 mM for β-D-galactosylpiperidine. In light of the preference of the cytosolic β-glucosidase for planar aryl aglycones, it was surprising to find that the piperidine compound is a more potent inhibitor relative to its planar pyridinium counterpart. The authors interpreted their findings as being consistent with the following mechanism. The base, β-D-galactosylpiperidine, deprotonates the AH group in the active site and the positive charge thus generated on the nitrogen atom in β-D-galactosylpiperidine could be stabilized by the carboxylate ion. This ion is usually involved in forming an ion pair with the glucosyl oxocarbonium ion. In contrast, the nitrogen in the β-D-galactosylpyridinium ion, which has already donated its lone pair of electrons, is incapable of functioning as a general base. Strong electrostatic interactions may account for the inhibitory potential of these nitrogenous compounds. In addition, the marked hydrophobic nature of the aglycone binding site (see below) would make it energetically unfavorable for a cationic derivative to partition from the bulk aqueous phase into the apolar interior of the enzyme.

Transglycosylation and Alcoholysis. Several observations have stimulated interest in the hydrophobic domains located in the vicinity of the active site of the cytosolic β-glucosidase: (i) the high affinity of the enzyme for hydrophobic affinity columns (eg., phenyl-Sepharose and octyl-Sepharose) (*19*), (ii) the marked inhibition of β-glucosidase activity by physiologically relevant amphipathic compounds such as glucosylsphingosine (*27*) and nonphysiologic amphiphilic alkyl

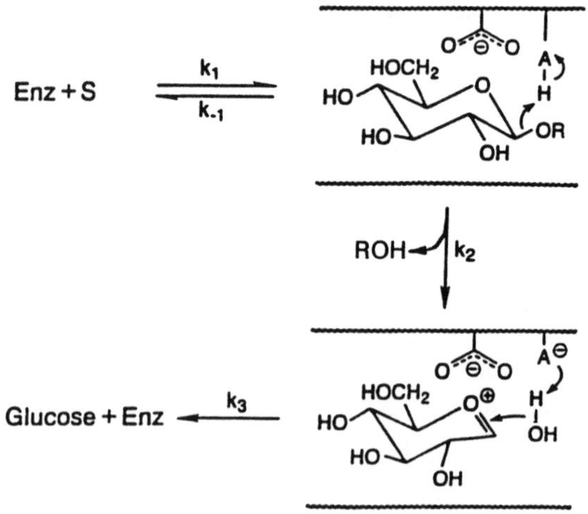

Figure 3. Postulated mechanism of action of cytosolic β-glucosidase. The participation of two critical amino acid residues in the catalytic process is indicated (See text for details). In the substrate, R refers to the aglycone moiety.

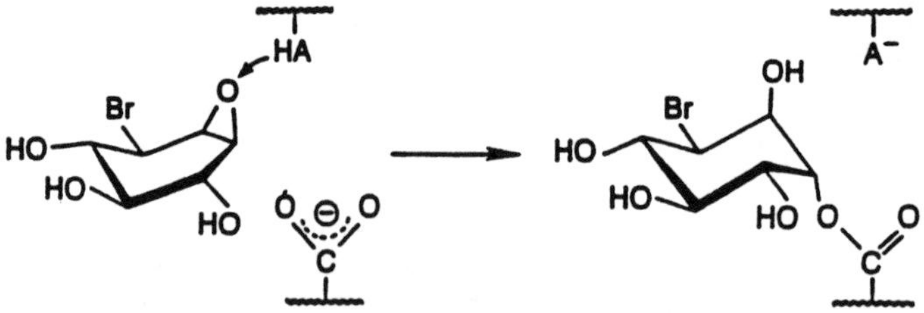

Figure 4. A postulated mechanism depicting the covalent modification of the critical catalytic residues in the active site of the β-glucosidase by BrCBE.

glycosides (*29*), and (iii) the modest stimulation of β-glucosidase activity by gangliosides such as GD1A (*40*). In an effort to learn more about these hydrophobic sites on the cytosolic β-glucosidase, we investigated the effects of alcohols on the kinetic properties of the guinea pig liver β-glucosidase.

Inclusion of increasing amounts of alcohols such as ethanol, n-propanol, n-butanol and n-pentanol in the β-glucosidase assay medium gives rise to biphasic activity versus alcohol concentration curves (Figure 5); at low concentrations, the alcohols stimulate β-glucosidase activity while at higher concentrations they inhibit the release of 4-MU from 4-MUGlc (*41*). Lineweaver-Burk analyses of this kinetic data also revealed a distinctly biphasic response to alcohols; at low concentrations, the n-alkanols increase the V_{max} 5 to 7-fold while at higher concentrations competitive inhibition is observed. With n-butanol, for example, increasing the concentration of the alcohol in the assay medium from 0 to 140 mM resulted in a progressive increase in V_{max} to a value 7-fold greater than the basal level without affecting the K_m. However, between 140 and 540 mM n-butanol, the K_m for 4-MUGlc increased from 0.14 to 0.93 mM (*14*). Secondary replots of 1/slope *versus* 1/[alcohol] were used to determine the binding constants (K_B values) for each alcohol (*42*). The linear nature of these secondary replots confirmed that a single molecule of alcohol was interacting with a hydrophobic site on the cytosolic β-glucosidase. Furthermore, the binding constants displayed a 4.0 to 4.5-fold decrease with the introduction of each additional carbon atom in the alcohol (*19*). The propensity to bind the longer chain alcohols was reflected in the K_B values of 555, 146, 34.1, and 7.47 mM for ethanol, n-propanol, n-butanol, and n-pentanol, respectively.

In contrast to n-butyl alcohol or isobutyl alcohol which are potent activators of the cytosolic β-glucosidase, compounds like sec-butyl alcohol, butylurea, and 1-butanesulfonic acid did not elicit a stimulatory response. Thus, the length of the alkyl chain, the shape of the molecule, and the presence of the hydroxyl group seem to be critical determinants of this interaction (*43*).

Alcohols are generally stronger nucleophiles than water in reactions involving nucleophilic substitutions. Therefore, in the solvolysis of the glycosyl-enzyme intermediate generated by any glycosidase, the rate of alcoholysis (k_4) will be greater than the rate of hydrolysis, k_3 (Figure 6). Indeed, on the basis of an analysis of the products formed when cytosolic β-glucosidase was incubated with various aryl-β-D-glucoside substrates in the presence of 200 mM n-butanol, we concluded that: i) the breakdown of these substrates proceeds through a common glucosyl enzyme intermediate, and ii) that the rate of alcoholysis (k_4) is 24-fold higher than the rate of hydrolysis (k_3) of the common intermediate.

The alcoholysis effect can increase the k_{cat} value only if the rate-limiting step in the cytosolic β-glucosidase catalyzed-cleavage of aryl-β-D-glucosides is the deglucosylation of the glucosyl-enzyme intermediate (k_3), i.e., if the rate of expulsion of the aglycone (k_2) is greater than k_3. The finding of an 8-fold increase in the rate of pNP (aglycone) release from pNPGlc elicited by n-butanol is therefore consistent with k_3 being the rate-determining step in the enzymatic hydrolysis of pNP-Glc in the absence of alcohols (*14*). However, when the amount of D-glucose product was quantitated we found that the rate of D-glucose release decreased as the concentration of n-butyl alcohol in the assay medium was

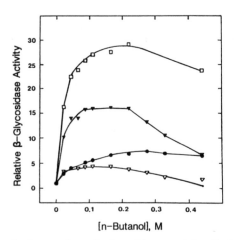

Figure 5. Effect of n-butanol on β-glycosidase activity. Relative activities of cytosolic β-glucosidase activity toward 4-MUGlc (●), 4-MUGal (△), 4-MUXyl (▼), and 4-MUAra (□) are plotted versus the concentration of n-butanol. (Reproduced with permission from reference 14. Copyright 1992 The American Society for Biochemistry and Molecular Biology.)

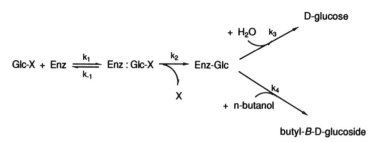

Figure 6. Scheme depicting alcoholysis catalyzed by the cytosolic β-glucosidase in the presence of n-butanol.
(Reproduced with permission from reference 14. Copyright 1992 The American Society for Biochemistry and Molecular Biology.)

increased. The latter observation is consistent with the hypothesis that the alcohol competes with water as a nucleophilic acceptor. Since $k_4 > k_3$ and $k_2 > k_3$, alcoholysis leads to a considerable increase in the overall reaction rate. Thin layer chromatographic analysis in our laboratory confirmed that the alcoholysis product, n-butyl-β-D-glucoside, was generated in a β-glucosidase assay performed in the presence of n-butanol. This finding also supports the postulate that the discrepancy between the expected and observed D-glucose yields in the presence of alcohols in the cytosolic β-glucosidase-catalyzed breakdown of aryl-β-D-glucosides is due to the formation of alcoholysis product, namely, an alkyl-β-D-glucoside.

The primary alcohol moieties of monosaccharides can also compete with water in the alcoholysis of β-glucosides; in fact, several glycosidases are known to catalyse transglycosylation to sugar acceptors (*44-46*). Distler and Jourdian demonstrated that a soluble aryl hexosidase from bovine liver is capable of catalyzing a transglycosylation reaction (*21*). We have shown that the guinea pig liver cytosolic β-glucosidase catalyzes the formation of pNP-β-D-gentobioside (β6→1) from pNP-β-D-Glc, and pNP-β-Gal(6→1)β-Gal from pNP-β-D-Gal. When the pNPGlc and pNPGal concentrations were 10 mM, gentobioside formation was about 14% of the amount of pNP release, and pNP-β-Gal(6→1)β-Gal formation occurred at 60% of the rate of pNP release. β-Glucosidase-catalyzed disaccharide formation was completely blocked by adding n-butanol (~0.4 M) to the reaction medium (*14*). It appears that the cytosolic β-glucosidase can accommodate relatively bulky aryl glycosides in the same site where linear alkanols normally bind during alcoholysis.

The effects of alcohols on cytosolic β-glucosidase-catalyzed disaccharide formation is accounted for very well by Segel's two-site model for a random bisubstrate rapid equilibrium system (*42*). According to this model, there are two hydrophobic sites in the catalytic center of the cytosolic β-glucosidase. One of the substrates (n-butanol) binds to a site designated B, but also competes for the binding of the second substrate (β-glucoside) to site A (Figure 7). The site designated B has the higher affinity for the alcohol, but alcohols can also interact at the A site, causing competitive inhibition. Affinity constants for interactions of alcohols at sites B and A can be designated as K_B and K_I, respectively. K_B and K_I values decreased about 4-fold for each additional methylene group in a series of linear alcohols (*19*), thus providing evidence for two hydrophobic sites at or near the enzyme's active center.

The relative affinity of pNPGal for site B on the β-glucosidase is less than the affinity for n-butanol. Therefore, disaccharide synthesis can occur in the absence of linear alcohols, as the B site is now available for occupancy by pNPGal. Separate K_m values can be measured for pNPGal binding to site A (0.46 mM for pNP release), and for binding to site B (6.5 mM for disaccharide product formation) (*14*). The potential biosynthetic implications of this type of glycosyl transferase activity remain to be explored (*47*).

Most of the β-glucosidase substrates contain a hydrophobic aglycone and a polar glycone moiety. BrCBE is known to interact with the site in the active center that binds the glycone moiety in the substrate. Our observation that n-butanol increases the K_i for BrCBE 4 to 5-fold indicates that n-butanol probably binds to the same domain on the guinea pig β-glucosidase at which BrCBE binds (Table II). This implies that the alkyl group in n-butanol could interact with the apolar site A while the hydroxyl group interacts, presumably through H-bonding, to the glycone binding site in the enzyme's active center.

In the nucleophilic attack of a carbonyl carbon by an alcohol, one would expect the reactivity of a family of alcohols (whose carbon number is invariant) to increase with the introduction of electron withdrawing groups that increase the acidity of the alcohols. We analyzed the effects of ethanol (pK_a = 16.0), ethylene glycol (pK_a = 15.0), 2-chloroethanol (pK_a = 14.3) and 2,2-dichloroethanol (pK_a = 13.5) and phenol (pK_a = 10.0) on cytosolic β-glucosidase activity in an effort to elucidate the nature of the nucleophilic species that attacks the glucosyl-enzyme

Table II. Effect of n-Butanol Addition on the Inhibition of Cytosolic β-Glucosidase by BrCBE

Addition	k_i/K_I $M^{-1}min^{-1}$	K_I mM	k_i min^{-1}
BrCBE	92.8	0.043 ± 12.9%	4.02 x 10^{-3} ± 2.8%
BrCBE plus n-butanol[1]	23.2	0.203 ± 12.8%	4.71 x 10^{-3} ± 4.4%

[1]n-Butanol was present at 0.4 % (v/v) in the assays.
Means ± SEM for the respective kinetic constants.

The following scheme indicates the mechanism of action of n-butanol in influencing the inhibition of guinea pig liver β-glucosidase by BrCBE, where E = β-glucosidase and I = BrCBE.

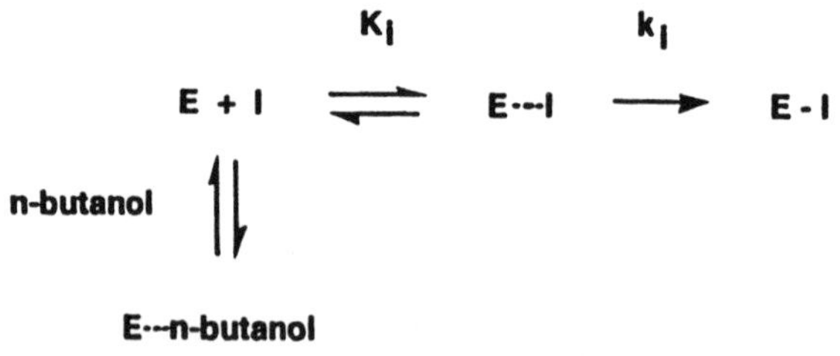

$$E + I \underset{}{\overset{K_I}{\rightleftharpoons}} E\text{---}I \xrightarrow{k_I} E\text{-}I$$

n-butanol

E---n-butanol

Note that the alcohol affects the affinity constant K_I, not the rate constant k_i. The second order rate constant of 92.8 M^{-1} min^{-1} agrees well with the value of 72.5 M^{-1} min^{-1} reported by Legler and Bieberich (29) for the inhibition of calf liver β-glucosidase by BrCBE.

intermediate (*72*). It is clear from the data presented in Figure 8, that for phenol and the two chloroethanols there was marked inhibition of β-glucosidase activity, even at concentrations as low as 0.05 M. In contrast, a 7-fold increase in the rate of 4-MU production was caused by adding 0.5 M ethanol or ethylene glycol to the assay medium. As stated previously, the ability of n-butanol to increase the overall rate of pNPGlc disappearance is due to the more rapid rate of alcoholysis relative to hydrolysis catalyzed by the cytosolic β-glucosidase (Figure 6). Therefore, the failure of phenol, 2,2-dichloroethanol and chloroethanol to increase the k_{cat} value for catalysis of 4-MUGlc hydrolysis is presumably due to their lower rates of solvolysis compared to water. These data led us to speculate that the hydroxyl oxygen of the alcohol gains a partial positive charge when proceeding from the free form in the aqueous medium to the transition state in the interior of the enzyme. Our findings indicate that increasing the acidity of the alcohols by inclusion of electron withdrawing groups in the carbon chain decreases the nucleophilic reactivity of the hydroxyl group toward the glucosyl oxocarbonium ion intermediate generated in the enzyme-catalysed reaction. These results are inconsistent with the hypothesis that the conjugate base of the alcohol serves as a reactive nucleophilic species in the transglucosylation reaction. The mechanism of formation of alkyl β-D-glucosides presumably involves a general base-catalysed removal of hydroxyl proton from the alcohol by a critical group in the active site of the β-glucosidase. A similar mechanism has been proposed by Greenzaid and Jencks for the case of alcoholysis of the acyl-esterase intermediate generated during the serine esterase-catalysed breakdown of acetate esters (*44*). In the case of cytosolic β-glucosidase, a more detailed analysis of the rates of alcoholysis of the various substituted ethanols is necessary in order to obtain a Brönsted plot of alcoholysis rate versus the pK_a of various nucleophilic reagents.

Substrate Specificity

A glycoside is formed when the hydroxyl group on the anomeric carbon atom (i.e., C(1) of aldoses or C(2) of ketoses) of the pyranose or the furanose of a monosaccharide is replaced by a molecule or group possessing a nucleophilic atom (*48*). The most common nucleophilic atom is the oxygen that is present in an alcohol or phenol. The resulting compound is an O-glycoside that consists of a glycone moiety linked to a nonsugar domain, hereafter referred to as the aglycone. This section which deals with the issue of substrate specificity will therefore be divided into two parts in order to address the significance of both the glycone and aglycone components of substrate molecules.

Glycone Specificity - Monosaccharides. The broad specificity of the cytosolic β-glycosidase is an interesting property of this enzyme (*24, 49*). The cytosolic enzyme is capable of catalyzing the hydrolysis of β-D-glucosides, β-D-galactosides, α-L-arabinosides, and β-D-xylosides. This list of substrates could be taken as evidence that the enzyme lacks specificity for the anomeric configuration or the saccharide moiety of its substrates. This, in fact, is not the case, since a major part of the glycone moiety is common to the four substrates. All four glycoside substrates possess the all *trans* equatorial configuration with respect to the oxygen

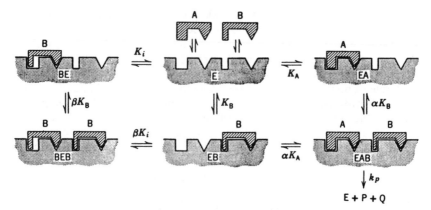

Figure 7. A two-site model for substrate inhibition in a rapid equilibrium system (Adapted from ref. 48). Substrate (pNPGal) usually has a higher affinity for site A, while alternate substrate B (n-butanol) has a higher affinity for site B, but substrates can bind to either site. Occupancy of site B by pNPGal leads to disaccharide synthesis, while occupancy of site A by n-butanol inhibits pNPGal hydrolysis.

(Reproduced with permission from reference 14. Copyright 1992 The American Society for Biochemistry and Molecular Biology.)

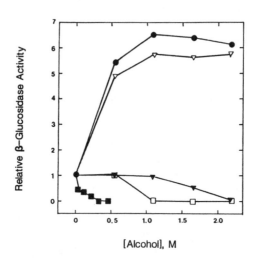

[Alcohol], M

Figure 8. The effect of electron withdrawing groups on the ability of alcohols to increase β-glucosidase activity. The relative activity of β-glucosidase for the hydrolysis of 4-MUGlc is plotted versus the concentration of ethanol (●), ethylene glycol (▽), 2-chloroethanol (▼) 2,2-dichloroethanol (□) and phenol (■).

atoms at positions C(1), C(2) and C(3) of the monosaccharide moiety (Figure 9). The structural variations occur only in the substituents at the C(4) and C(5) atoms of the pyranose ring.

The order of catalytic efficiency towards the methylumbelliferyl (MU) glycosides is MUGlc > MUGal = MUAra > MUXyl (*19, 26, 29*). Estimation of K_m and V_{max} values for the various glycosides attached to a common aglycone have provided some insight into their varying rates of hydrolysis. A three-point attachment model was first proposed by Daniels and associates for the saccharide binding or recognition domain in order to account for the varying catalytic efficiencies displayed by the cytosolic β-glucosidase towards the different glycoside substrates (*29*). This model was later refined by LaMarco and Glew (*50*). As is evident in Figure 10, all four of these substrates possess an equatorial bond at the anomeric carbon atom. This absolute requirement may indicate that this equatorial anomeric carbon interacts with some critical residue in the enzyme's active site. It may also indicate a need for the glycoside to be protonated by some rigid proton donor, AH (Figure 3), in the active site. The latter step would facilitate expulsion of the aglycone. Since β-D-glucosides are preferred over their 4-epimer counterparts, namely β-D-galactosides, it was inferred that the glycohydrolase favors substrates that possess an equatorial group at the C-4 position. Furthermore, the finding that MUXyl (which differs from MUGlc in that it possesses a hydrogen atom in place of the CH_2OH group at the C(5) position) was the poorest substrate amongst the four glycosides mentioned above, led Daniels et al. to propose a model involving groups on the enzyme that could specifically recognize the C(1), the C(4) and the C(5) substituents in the pyranose ring (*29*).

In data derived from a number of studies which were performed with cytosolic β-glucosidases from a variety of mammalian species and tissues, there is general agreement that the K_m and V_{max} values for the pentoside substrates are considerably lower than those of their hexoside counterparts (*19, 26, 49*). Chester and associates pointed out that human liver β-glucosidase has a greater affinity for the smaller substrates, MUXyl and MUAra, and that this high affinity slowed the rate of hydrolysis of these pentosides (*49*). Based on the higher affinity of the β-glucosidase for pentoside substrates MUXyl and MUAra, Pocsi and Kiss recently proposed that the recognition of substituents at C(4) and C(5) in the glycone moiety of the glycoside substrates is unlikely to result in stronger binding of the substrates to the glycone-interacting subsite in the catalytic center of the β-glucosidase (*26*). The lack of a C(6) primary hydroxyl group in the pentoside substrates, taken together with the higher affinity of the pentosides for the active site of the cytosolic β-glucosidase, validates this hypothesis. Pocsi and Kiss (*26*) also speculated that the active site of the mammalian cytosolic β-glucosidase possesses bulky groups at these particular recognition points and that they could compress the glycoside molecule, thereby deforming the pyranoside ring.

This substrate distortion hypothesis provides a thermodynamic rationale that could account for the varying catalytic efficiencies exhibited by the cytosolic β-glucosidase towards different glycosides. This hypothesis implies that the enzyme uses the C(5) primary hydroxyl group in β-D-glucosides and β-D-galactosides as a lever to distort the pyranose ring into the half-chair conformation. The significance of the half-chair conformation was discussed above. It is important to appreciate that the resulting ES complex is destabilized and raised to a higher energy level relative to both the free enzyme and substrate. Destabilization of the ES complex, in effect, translates into a lowering of the activation energy for the hydrolytic reaction and an increased rate of hydrolysis (51). The absence of such an interaction, as is the situation with β-D-xylosides and α-L-arabinosides, could account for the stronger binding and lower rates of hydrolysis observed with these pentosides. It is noteworthy that the catalytic mechanism proposed for lysozyme that is based on X-ray studies of the protein's structure includes substrate deformation as a prerequisite for substrate binding to the active site (1, 31).

Differences in the K_m values of cytosolic β-glucosidase for various glycoside substrates would normally be assumed to be based on changes in the intrinsic affinity of the enzyme's active site for the glycone moiety, namely the K_s value. Alternatively, since $K_m = K_s (k_3/k_2+k_3)$, the variations in K_m values for these glycoside substrates could arise from differences in their rates of glycosylation (k_2) and deglycosylation (k_3) in the absence of significant differences in the K_s values of the four glycoside substrates. For example, if the K_s value were the same for 4-MUGlc, 4-MUGal, 4MUXyl and 4-MUAra, and if the k_2/k_3 ratio were to display a decrease in the order 4-MUAra > 4-MUXyl > 4-MUGlc > 4-MUGal it follows that the K_m values would increase in the order 4-MUAra < 4-MUXyl < 4-MUGlc < 4-MUGal.

The effects of n-butanol on the β-glucosidase-catalysed cleavage of 4-MUGlc, 4-MUGal, 4-MUXyl and 4-MUAra have been used to inquire whether K_m differences arise from variations in the intrinsic affinity of the β-glucosidase for different glycones, or from differences in glycosylation (k_2) and deglycosylation (k_3) rates (14). One can use the approach described by Fersht to obtain estimates of glycosylation (k_2) and deglycosylation (k_3) rate relationships (52). For example, when using an aryl-β-D-glucoside like pNPGlc or MUGlc as substrate for the β-glucosidase, the ratio of the rates of alcoholysis/ hydrolysis (i.e., k_4/k_3) is 24:1 (14). The K_m values for MUGlc, MUGal, MUXyl and MUAra are 0.15 mM, 0.48 mM, 0.040 mM and 0.08 mM, respectively (14). If k_{cat+} in the presence of n-butanol is $k_2(k_3 + k_4)/(k_2 + k_3 + k_4)$ and k_{cat-} in the absence of n-butanol is $k_2k_3/(k_2 + k_3)$, knowing that $k_4/k_3 = 24$ and $k_{cat+}/k_{cat-} = 8$ (since n-butanol causes an 8-fold increase in the k_{cat} for the breakdown of pNP-Glc), one can derive the relationship $k_2 = 10.3 k_3$. If one then uses the relationship $K_m = K_s(k_3/k_2+k_3)$, and a K_m value of 0.15 mM for MUGlc, the estimated K_s value for the interaction of 4-MUGlc and the cytosolic β-glucosidase is 1.55 mM. On the basis of the K_m values for these four different substrates (14), and assuming that the K_s value is the same for all four of the 4-methylumbelliferyl glycosides, we calculated the k_2/k_3 ratios for the breakdown of 4-MUGal, 4-MUAra, 4-MUXyl to be 2.5, 38 and 18, respectively. These theoretical ratios are in reasonable agreement with the actual 4-, 30- and 15-fold increases in k_{cat} values observed for the breakdown of 4-MUGal, 4-MUAra and 4-MUXyl, respectively, by the cytosolic β-glucosidase when n-butanol is included in the reaction medium (Figure 5). The discrepancies could be rationalized on the basis of different rates of alcoholysis of the various enzyme-glycosyl intermediates. It appears, therefore, that the differences in K_m values for the substrates 4-MUGlc, 4-MUGal, 4-MUXyl and 4-MUAra are related to variations in the kinetic rate constants for the hydrolysis of these glycosides by the cytosolic β-glucosidase, and are not due to subtle differences in the affinity of the enzyme for these substrates.

This contention that the active site of the cytosolic β-glucosidase has a similar affinity for substrates with different glycone moieties is supported by the results of two other studies (*14, 26*). Since $K_m = K_s(k_3/k_2 + k_3)$, the K_m values of the cytosolic β-glucosidase for different substrates could be influenced by differences in the rates of glucosylation (k_2) and deglucosylation (k_3). If this postulate is correct, one would expect the K_s values to be identical for substrates that possess the same aglycone but different glycone moieties, and which exhibit similar rates of glycosylation (k_2) and deglycosylation (k_3). Competitive inhibitors of the cytosolic β-glucosidase would be ideal reagents for verifying this hypothesis due to their extremely low k_2 values, and they would be expected to exhibit identical K_i values regardless of their glycone moieties. This contention is based on the circumstance in which k_2 is very low ($k_2 <<< k_3$), and $K_m = K_s(k_3/k_2 + k_3)$ reduces to $K_m = K_s$. This expected behavior is confirmed by at least two recent reports. First, on the basis of their studies of the inhibition of the pig kidney cytosolic β-glucosidase by thiophenyl and by p-aminothiophenyl glycosides, Pocsi and Kiss reported that the K_i values obtained were independent of the carbohydrate moiety of the inhibitors under study (*26*). For instance, the K_i values for thiophenyl-β-D-glucoside and thiophenyl-β-D-galactoside for the pig kidney cytosolic β-glucosidase were 840 μM and 670 μM, respectively. Second, we reported K_i values of 10 μM and 9.9 μM for the inhibition of human liver cytosolic β-glucosidase caused by the reversible, competitive inhibitors octyl-β-D-glucoside and octyl-β-D-galactoside, respectively (*41*). Neither the thiophenyl-nor the alkyl-glycosides are hydrolysed to any significant extent by the cytosolic β-glucosidase (*26, 41*).

Glycone specificity - disaccharides. We found disaccharides to be very poor substrates of the cytosolic β-glucosidase. Little or no hydrolysis occurred when the human liver β-glucosidase was incubated with lactose, lactulose, cellobiose, 3-O-β-D-galactosyl-D-arabinoside or gentiobiose (*21, 24*). Furthermore, the enzyme does not catalyze the hydrolysis of glycosidic linkages in the polysaccharides xylan (Glc β3→1 Glc) and laminarin. However, the effectiveness of alkyl-β-D-glucosides as inhibitors of the cytosolic β-glucosidase indicates that the enzyme probably possesses a hydrophobic subsite in its catalytic center (*41*). We reasoned, therefore, that disaccharides such as cellobiose, lactose or gentiobiose are poor substrates in the β-glucosidase assay because they lack a hydrophobic aglycone moiety. Accordingly, glycosides that contain disaccharide moieties terminating in a β-linked glucose or galactose unit and which possess a hydrophobic aryl aglycone (e.g., p-nitrophenol) should be substrates for the cytosolic β-glucosidase. This idea was pursued using p-nitrophenyl disaccharides as substrates for the cytosolic β-glucosidase.

When the p-nitrophenyl derivatives of cellobioside, gentiobioside, lactoside and galactosyl-β(1→6)-galactopyranoside were tested as substrates, only pNP-gentiobioside was cleaved at an appreciable rate by the guinea pig liver enzyme (*15*). These findings indicate that the cytosolic β-glucosidase prefers substrates which contain a β1→6 linkage between the two sugar moieties and that the enzyme will hydrolyze a β-D-glucosyl residue, but not a β-D-galactosyl residue at the nonreducing end of disaccharide glucosides. It is noteworthy that human glucocerebrosidase does not catalyze the hydrolysis of pNP-gentiobioside or any of the above-mentioned disaccharides (data not shown).

It is interesting that the extent of hydrolysis of pNP-gentiobioside is at least 30-fold higher than that of pNP-cellobioside or pNP-lactobioside (*15*). This observation provides additional insights into the mechanism of action of cytosolic β-glucosidase. It is likely that the first step in the enzyme-catalysed hydrolysis of glycosides involves the protonation of the anomeric oxygen by an acid in the active site of the enzyme, and concomitant expulsion of the aglycone (Figure 3). Substantial nucleophilic assistance to C(1)-O-R bond breaking may be essential in order to get efficient acid catalysis by the proton donor AH in the active site of the cytosolic β-glucosidase. Kirby has pointed out two likely sources for such nucleophilic assistance: first, the lone pair of electrons in the pyranosyl ring oxygen, and second, the carboxylate group that is believed to participate in stabilization of the oxocarbonium ion or covalent glucosyl-enzyme intermediate (*53*). For instance, to expel the terminal glucose moiety from pNP-gentiobioside, the lone pair of electrons in the ring oxygen of the terminal glucose moiety may provide nucleophilic assistance to breaking the β1→6 linkage between the two glucose units.

The donation of a lone pair of electrons by the ring oxygen to C(1)-O-R bond breaking would be potentiated by the presence of a negative charge proximal to the ring oxygen. Alternatively, the committment of this lone pair of electrons to a H-bond will clearly make nucleophilic assistance from the ring oxygen less likely in the mechanism proposed above. Cellulose and lactose exist mainly in the form of extended, drawn-out linear chains (*54*). Both disaccharides are stabilized by a hydrogen bond between the hemiacetal ring oxygen and the oxygen at C(3) of the neighboring saccharide moiety (Figure 10). The absence of such an intramolecular H-bond in pNP-gentiobioside may contribute to the increased nucleophilicity of its ring oxygen; this could be the underlying basis for the higher rate of hydrolysis of pNP-gentiobioside compared to pNP-cellobioside by the guinea pig liver cytosolic β-glucosidase.

Aglycone Specificity. Data presented by Legler and Bieberich for the rates of hydrolysis of a family of alkyl-β-D-glucosides by the calf liver cytosolic β-glucosidase illustrates why it is not always favorable for a substrate to bind too strongly to an enzyme (*25*). They showed that the K_m values diminish as the chain length of the alkyl group of alkyl-β-D-glucosides increases. For instance, the K_m values for butyl-β-D-glucoside and octyl-β-D-glucoside interacting with the calf liver cytosolic β-glucosidase are 280 μM and 1.9 μM, respectively, while the corresponding k_{cat} values are 31 min^{-1} and 11 min^{-1}, respectively. Apart from the fact they are poor leaving groups, the aglycone groups of the alkyl-β-D-glucosides bind with such a high affinity to the hydrophobic interior of the enzyme that it is likely they force the enzyme reaction into a thermodynamic trough or free energy

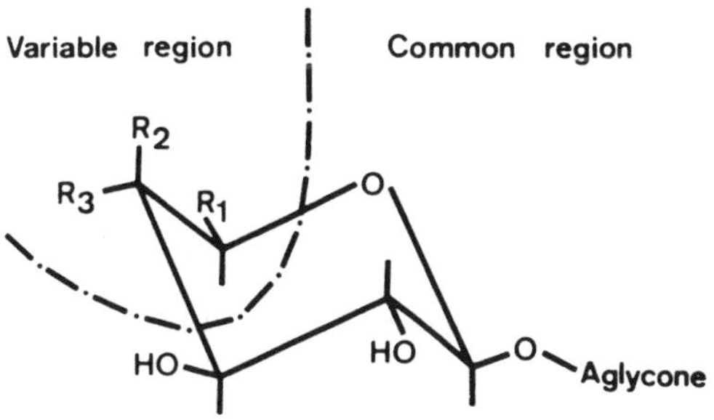

	R_1	R_2	R_3
ß-D-galactoside	$-CH_2OH$	$-OH$	$-H$
ß-D-glucoside	$-CH_2OH$	$-H$	$-OH$
ß-D-fucoside	$-CH_3$	$-OH$	$-H$
ß-D-xyloside	$-H$	$-H$	$-OH$
α-L-arabinoside	$-H$	$-OH$	$-H$

Figure 9. An illustration of the common and variable regions of the glycoside substrates of the cytosolic β-glucosidase.
(Reproduced with permission from reference 49. Copyright 1976 Elsevier Science Publishers BV.)

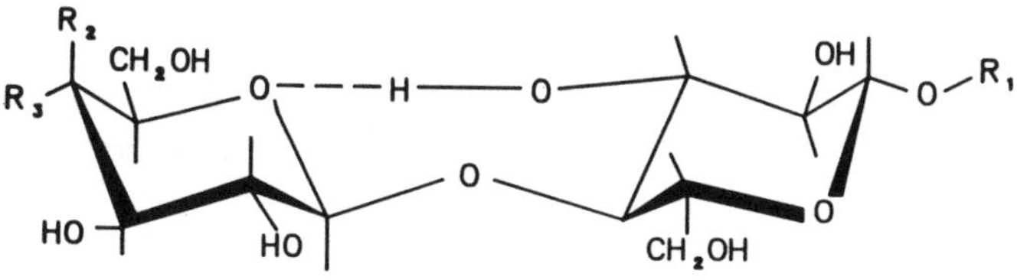

pNP—Disaccharide	R_1	R_2	R_3
1) pNP—Cellobioside	pNP	H	OH
2) pNP—Lactoside	pNP	OH	H

Figure 10. The intramolecular hydrogen bonding that may occur between the ring oxygen and the C(3)-OH in the neighboring sugar moiety in both pNP-β-D-cellobioside and pNP-β-D-lactoside (Adapted from ref. 55).

minimum, thereby making alkyl-β-D-glucosides very poor substrates for the cytosolic β-glucosidase. Legler and Bieberich envision the aglycone binding site of the enzyme as a narrow cleft which prefers rigid, planar, hydrophobic moieties to long, flexible alkyl chains (25).

According to the catalytic mechanism proposed in Figure 3, in order for a glucoside to be a good substrate its aglycone moiety must possess powerful electron withdrawing substituents. However, the hydrophobic nature of the aglycone binding subsite in the catalytic center of the β-glucosidase will tend to exclude polar substituents attached to the aglycone moiety of the substrate. It is important to appreciate the fact that inclusion of polar electron-withdrawing groups that increase the V_{max} would also be expected to increase the K_m of the enzyme for such substrates. Noteworthy is the fact that two widely used nonphysiologic β-glucosidase substrates, namely p-nitrophenyl-β-D-glucoside and 4-methylumbelliferyl-β-D-glucoside, have an aryl, hydrophobic aglycone moiety as a common feature. A planar and aryl character of the aglycone moieties has also been a good predictor for naturally occurring glycosides (e.g., prunasin) that have been shown to be good substrates for the cytosolic β-glucosidase (43).

Physiologic Substrates and Functions

Glycosphingolipid degradation. The recognition that lysosomal storage disorders are the result of deficiencies of acid pH optima hydrolases has led some investigators to speculate that the cytosolic enzyme might play a protective role in certain lipid storage disorders by providing an alternate metabolic route for the degradation of accumulated substrate. In the case of Gaucher disease the neutral pH optimum, soluble β-glucosidase could conceivably play a role in alleviating the disease by degrading accumulated glucocerebroside or glucosylsphingosine products of the degradation of glucocerebroside. Similarly, if hydrolysis of the β-galactosidic linkage at the non-reducing end of ganglioside GM1 by the broad-specificity cytosolic β-glucosidase were possible, this would reduce the lipid storage in GM1 gangliosidosis. In spite of these intriguing possibilities, Daniels and associates could not find assay conditions under which the cytosolic β-glucosidase would exhibit significant glucocerebrosidase activity (8, 9). Distler and Jourdian reported that the bovine liver aryl hexosidase does not cleave glucocerebroside, lactocerebroside, galactocerebroside or GM1 ganglioside (21). Recently, however, Legler and Bieberich demonstrated that at pH 5.5 the calf liver soluble β-glucosidase catalyzed the hydrolysis of glucosylsphingosine with a turnover number that was 4% that of the nonphysiologic substrate, p-nitrophenyl-β-D-glucoside (25). However, the glucosylsphingosine hydrolysis they observed *in vitro* was accompanied by strong product (sphingosine) inhibition.

Glucosylsphingosine and galactosylsphingosine are potent inhibitors of the hydrolysis of pNPGlc by human liver broad-specificity β-glucosidase (50). These toxic lipoidal amines accumulate in the brains of patients affected with Gaucher or Krabbe disease (55). While there is insufficient evidence that a soluble, broad-specificity β-glucosidase is present in the brain, the inhibition of a closely related nonlysosomal membrane-associated isoform of the cytosolic enzyme could contribute to the pathophysiology of these disorders. We have found that guinea pig brain contains a membrane-bound, broad-specificity β-glucosidase that cross-reacts with antibodies directed against the β-glucosidase isolated from guinea pig liver cytosol (R. H. Glew, unpublished observation).

In spite of the evidence that glucocerebroside and ganglioside GM1 are poor substrates of the cytosolic β-glucosidase, there has been speculation that a combined deficiency of the cytosolic β-glucosidase and certain lysosomal hydrolases might add to the severity of the clinical phenotype of particular lysosomal storage disorders. Tissue extracts of several patients affected with the more severe type I variant of GM1 gangliosidosis were characterized by a dual deficiency of both the lysosomal acid β-galactosidase and the broad-specificity cytosolic β-glucosidase/β-galactosidase (*56*). However, in patients with the milder type II form of GM1 gangliosidosis, normal levels of the cytosolic β-galactosidase were observed. A similar finding was reported in a study performed with liver obtained at the time of autopsy from patients who had died of type I and type II Gaucher disease. Two different laboratories observed independently that some cases of the most severe neurologic phenotype of Gaucher disease exhibited a dual deficiency of lysosomal glucocerebrosidase and broad-specificity cytosolic β-glucosidase activities (*8, 57*). In contrast, in patients with the type I, adult, nonneurologic form of the disease the tissue levels of the cytosolic β-glucosidase were normal or elevated. The factors that contribute to the different clinical phenotypes in Gaucher disease remain obscure.

More extensive surveys are needed before any conclusive link can be established between the expression of the cytosolic β-glucosidase activity and a benign phenotype in these glycolipid storage disorders. The latter issue could prove contentious because of the low probability of a dual deficiency within one inherited disorder and of the inability of the cytosolic β-glucosidase to catalyze the hydrolysis of glucosylceramide. This question is likely to be resolved when the cDNA that codes for the cytosolic β-glucosidase is cloned and sequenced. The finding of structural or regulatory mutations in patients with the most severe cases of Gaucher disease or GM1 gangliosidosis would point to a role for the cytosolic β-glucosidase in the etiology of these disorders.

Catabolism of Mucopolysaccharides and Glycoproteins. The search for a physiologic function that would implicate the cytosolic β-glucosidase as a determinant in the clinical severity and pathophysiology of Gaucher disease has been an important motivation for investigators seeking to identify the enzyme's natural substrates. The enzyme has been hypothesized to play a role in the metabolism of mucopolysaccharides. Since Gaucher fibroblasts have been shown to contain increased quantities of mucopolysaccharides (*58*), and because mucopolysaccharides can inhibit glucocerebrosidase activity (*63*), it was proposed that a reduction in cytosolic β-glucosidase activity might exacerbate the glucocerebrosidase deficiency observed in patients with Gaucher disease. However, research over the past 25 years has failed to identify any role for the enzyme in the catabolism of mucopolysaccharides. Distler and Jourdian reported that keratan sulfate was not a substrate for the bovine liver aryl hexosidase (*21*). The inability of the enzyme to hydrolyze xylosyl-O-serine led to rejection of the hypothesis that the enzyme's β-xylosidase activity might function to cleave xylosyl-O-serine linkages in chondroitin sulfate and other mucopolysaccharide-protein complexes (24, 60). In fact, this finding is consistent with the observation that xylosyl-O-serine is found in the human urine and is not subject to metabolism to its amino acid and sugar constituents by a β-xylosidase prior to excretion (61).

The catalytic flexibility of the cytosolic β-glucosidase has also been an attribute that has strongly influenced the direction of research with regard to identifying its physiologic substrates. With the knowledge that many glycoproteins possess β-galactosyl residues in the peripheral regions of their oligosaccharide side chains, we tested [^{14}C]-galactosyl- ovalbumin as a substrate for the cytosolic enzyme (24). The glycosidase did not remove detectable quantities of the β-linked terminal galactose residues from the glycoprotein. A similar result was obtained by Distler and Jourdian when they inquired if desialyzed, galactose-terminated α_1-acid glycoprotein was a substrate for this enzyme (21). We postulated that the cytosolic enzyme may metabolize glycofucopeptides present in mammalian tissues. Our preliminary studies with the guinea pig liver cytosolic β-glucosidase have revealed that it did not degrade the β-glycosyl linkages in glucosyl β 1→3 fucosyl α 1→ threonine (or N-acetylated threonine) and glucosyl β1→3fucosylα1→serine (R. H. Glew, unpublished observation). Collectively, these findings do not support a role for the enzyme in the degradation of the oligosaccharide chains of glycoproteins.

Hydrolysis of Steroid Glucosides. Injection of 17-α-estradiol-3-β-D-glucoside and 17-β-estradiol-3-β-D-glucoside into rabbits and humans led to the production of the same glucuronidated metabolites as those generated by the injection of the free steroid (62, 63). It was concluded that hydrolysis of the the glycosidic bonds of the steroid glucoside must have occurred *in vivo* in order to permit formation of the respective 3-glucuronides. The 3-glucosides were absent in rabbit excreta despite their demonstrated formation in rabbit liver.   These results strongly indicated that tissue β-glucosidases act to expose the 3-hydroxyl group of steroid glucosides prior to their conjugation with glucuronic acid. A steroid glucosidase activity was identified soon thereafter in rabbit liver cytosol and the enzyme was purified to homogeneity (64, 65). The rabbit liver β-glucosidase exhibited low turnover numbers with 17-α-estradiol-3-β-D-glucoside, 17-β-estradiol-3-β-D-glucoside,   estrone-3-β-D-glucoside   and   deoxycorticosterone-β-D-glucoside substrates. We demonstrated that the same steroid glucoside-cleaving activity was possessed by the human liver β-glucosidase; however, we found that steroid β-D-glucosides were hydrolysed at only about one percent the rate at which 4-MUGlc was hydrolyzed (8).

Low turnover numbers have been taken as evidence that steroid glucosides are not physiologically significant substrates for the soluble β-glucosidase. This position must be reconciled with the evidence that exogenous steroid glucosides were apparently hydrolysed completely *in vivo*. However, the overall importance of such a reaction may be minimal in the absence of a significant endogenous source of steroid glucosides.

Transport of Glucosides. The sensitivity of the kidney enzyme to inhibition by phloridzin (60), taken together with its localization in the proximal convoluted tubular cells of the kidney (66), led to the hypothesis that the soluble β-glucosidase may be involved in renal reabsorption of sugars. The high content of cytosolic β-glucosidase activity in rat kidney and intestine (*24*) the tissues with the greatest capacity for transport does point to a possible role for the enzyme in glucoside transport. We speculated initially that the enzyme may function in the cytosol of the enterocyte to release glucose from glucosides absorbed from the diet (*24*). However, in a more recent study of plant glucoside transport, we found that certain plant glucosides which are β-glucosidase substrates (e.g., vicine, amygdalin, and prunasin) escape hydrolysis during passage across the guinea pig ileum (*15*).

A Biotransformation Role Toward Selected Plant β-Glucosides. Lysosomal glucocerebrosidase has a specific requirement for glucose as the glycone moiety of its substrates. We postulated that the cytosolic β-glucosidase's lack of strict substrate specificity may have special functional significance. A lack of specificity could contribute to a wider role for the enzyme in hydrolyzing a broad range of naturally occurring dietary glycosides.

LaMarco and Glew reported that L-picein, a plant phenolic glucoside, was efficiently hydrolyzed by the guinea pig liver cytosolic β-glucosidase (*19*). We recently determined the rates of hydrolysis of several other plant glucosides that could conceivably find their way into the diets of man and animals (*15*). It is well documented that cyanide poisoning can occur in mammals subsequent to hydrolysis of ingested cyanogenic plant glucosides (*67*). Also, phenolic glucosides have been found in nearly every higher plant species that has been examined for the presence of these compounds (*68*). We investigated the possibility that some of these phenolic glucosides or cyanogenic glucosides would be hydrolysed by the cytosolic β-glucosidase. Arbutin was hydrolysed to 60%, and salicin to 30% of the rate at which pNP-Glc was hydrolyzed by the guinea pig liver cytosolic β-glucosidase (*15*). Of the cyanogenic glucosides tested, D-amygdalin was hydrolyzed most rapidly, at 35% of the rate at which pNP-β-glucoside was hydrolyzed by the cytosolic β-glucosidase. Linamarin and neolinustatin were also hydrolyzed, but at relatively low rates. The pyrimidine glucoside, vicine, the causative agent in the hemolytic anemia of favism (*69, 70*), was hydrolyzed by the cytosolic β-glucosidase at about 15% the rate of 4-MUGlc hydrolysis (*15*).

We next asked: Does the cytosolic enzyme act as an endo- or exo-glucosidase? The endolytic and exolytic pathways of glycosidic cleavage of D-amygdalin would generate distinct product patterns during the course of the enzymatic reaction. Endolytic action on D-amygdalin would produce gentiobiose and the aglycone, mandelonitrile as products, whereas exolytic cleavage would yield the monosaccharide glucoside, prunasin (mandelonitrile-β-Glc), and the aglycone, mandelonitrile. The monosaccharide intermediate, prunasin, was identified by reversed-phase HPLC as the product of D-amygdalin hydrolysis by the cytosolic β-glucosidase, leading to the conclusion that the enzyme is an exo-β-glucosidase (*15*).

The cytosolic β-glucosidase could play a role in the biotransformation of dietary glucosides if these molecules gain access to the liver. We have used the everted gut sac technique (*71*) to show that prunasin, D-amygdalin, and vicine are all transported rapidly across the guinea pig ileum without appreciable hydrolysis (15). Thus, these plant glucosides could be pathophysiologically relevant substrates of the cytosolic β-glucosidase of mammalian liver. These results also indicate that it may be possible to target drugs to the liver by making inactive β-D-gentiobioside or β-D-glucoside derivatives. Efficient transport of the glycosides to the liver, and subsequent hydrolysis by hepatic cytosolic β-glucosidase could make drugs available to the liver in an active form.

Future Studies

Unquestionably, the most fruitful effort at the moment would be to isolate and sequence the gene for the cytosolic β-glucosidase. Having the cDNA sequences would permit not only comparisons with other β-glucosidases, particularly the lysosomal β-glucosidase, but also provide molecular tools for probing the structure and function of the enzyme. The availability of the β-glucosidase cDNA permits one to carry out site-directed mutagenesis, and to produce the enzyme in high-yield expression systems. Site-directed mutagenesis of the cDNA could provide a means for testing the hypothesis that the cytosolic β-glucosidase possesses distinct binding sites for glucosides and alcohols. Based on our two-site model of the cytosolic β-glucosidase (Figure 7), it is conceivable that altering the apolar nature of the high affinity alcohol binding site should abolish the stimulatory effect of alcohols on the rate of substrate cleavage, and eliminate the ability of the β-glucosidase to catalyze transglucosylation to alkyl-alcohol acceptors (e.g., n-butanol) or to glycosides such as pNP-β-D-glucose and pNP-β-D-galactose. Site-directed mutagenesis of the cDNA would also provide a means of identifying those amino acid residues involved in the catalytic mechanism.

We have used polyclonal antibody specific for the guinea pig liver β-glucosidase to isolate several clones from a λgt11 guinea pig liver cDNA library. The largest clone appears to carry a full-length (2.1 kbp) insert coding for the complete guinea pig liver cytosolic β-glucosidase. The inserts have been amplified by PCR, using primers specific for the regions 5' to the EcoRI site of λgt11, and cloned into M13mp19 for DNA sequencing. The largest insert has been labelled and is being used as a probe to screen a human liver λgt11 cDNA library and isolate the human cytosolic β-glucosidase cDNA.

A second area of research that is ripe for exploitation involves the study of the pathophysiologically relevant plant glucoside substrates. In demonstrating that cyanogenic glucosides, the hemolytic agent vicine, and the neurotoxic cycasin of the cycad seed are good substrates of the broad-specificity β-glucosidase of mammalian liver, we are probably only touching the surface of a large number of glycosylated compounds in the plant kingdom that are substrates for the cytosolic β-glucosidase.

The β-glucosidase-catalyzed cleavage of some of these plant glycosides could give rise to hitherto unrecognized toxic aglycone moieties which cause disease in man and animals. It would be useful, therefore, to expand the search for plant glycosides that are degraded by the mammalian cytosolic β-glucosidase. In addition, there is a need for animal studies that would clarify the role of this enzyme in the pathophysiology of cycasin, cyanogenic glycosides, and vicine toxicity. These studies should address the following questions: 1) What are the rates and mechanisms of transport of specific plant glycosides across the intestine? 2) How do sustrates such as vicine and prunasin escape hydrolysis by the intestinal β-glucosidase? 3) Which organs are capable of extracting these same plant glucosides from the blood? 4) Which organ's β-glucosidase is responsible for the hydrolysis of various ingested plant glycosides? For example, is it the cytosolic β-glucosidase of liver or brain that degrades cycasin? A spin-off of such studies would be the elucidation of the substrate specificity of the cytosolic β-glucosidase.

The transglycosylation phenomenon is another aspect of the cytosolic β-glucosidase that merits further study. There is need for additional information regarding the mechanism of the transglycosylation reaction between donor substrates (e.g., pNP-β-D-galactose, pNP-β-D-glucose) and acceptor alcohols (e.g., n-butanol, n-octanol). There are other specific questions that also need to be addressed: 1) What are the donor and acceptor specificities of substrates in the transglycosylation reaction with respect to both the sugar and the aglycone domains? 2) What is the maximum size of the oligosaccharide chain of the transglycosylation product that can be generated by the cytosolic β-glucosidase? Can it form glycosides having tri-, tetra- and pentasaccharide chains? Another interesting question is why is it that the specificity of the enzyme for disaccharide glycoside substrates is much narrower than that for monosaccharide glycosides. For example, the guinea pig β-glucosidase hydrolyzes pNP-β-D-galactose, pNP-β-D-glucose and pNP-β-D-glucosyl(6→1)glucose, but not pNP-β-D-glucosyl(6→1)galactose.

Another related issue is the question of whether primary alcohols present in mammalian tissues can serve as sugar acceptors in the cytosolic β-glucosidase-catalysed transglycosylation reaction. Specifically, if one incubates pNP-β-D-glucose and the guinea pig β-glucosidase with retinol or dolichol, is it possible to generate retinyl- or dolichyl-glycosides? Having raised this question for the cytosolic β-glucosidase, one is then led to ask if the lysosomal counterpart, glucocerebrosidase, is also capable of catalyzing transglycosylation reactions. Since tissues of most patients with Gaucher disease contain substantial amounts (15 to 30% of normal) of residual glucocerebrosidase, and enormous concentrations of storage lipid (e.g., glucocerebroside) it would be interesting to see if glucocerebrosidase catalyzes the transfer of glucose from pNP-β-D-glucose or glucocerebroside to hydrophobic alcohols.

Literature Cited

1. Walsh, C.T. In " Enzymatic Reaction Mechanisms "; Freeman, San Francisco, CA. 1979, 267-307.
2. Price, R.G.; Dance, N. *Biochem. J.* **1967**, *105*, 877-883.
3. Patel, V.; Tappel, A.L. *Biochim. Biophys. Acta.* **1969**, *192*, 653-662.
4. Glew, R.H.; Basu, A.; LaMarco, K.L; Prence, E.M. *Lab. Invest.* **1988**, *58*, 5-25.
5. Brady, R.O.; Kanfer, J.N.; Shapiro, D. *Biochem. Biophys. Res. Commun.* **1965**, *18*, 221 -225.
6. Patrick, A.D. *Biochem. J.* **1965**, *97*, 17c-18c.

7.   Grabowski, G.A.; Gatt, S.; Horowitz, M. *Crit. Rev. Biochem. Mol. Biol.* **1990**, *25*, 385-414.
8.   Daniels, L.B.; Coyle, P.J.;Chiao, Y-B.;Glew, R.H.; Labow, R.S. *J.Biol.Chem.* **1981**, *256*, 13004-13013.
9.   Daniels, L.B.; Gnarra, J.R.; Glew, R.H. In "Gaucher Disease: A Century of Delineation and Research "; Grabowski, G.A., Desnick, R.J., and Gatt, S., Eds.; Alan Liss, New York, NY, 1982, 333-355.
10. Robinson, D.S.; Price, R.G.; Dance, N. *Biochem. J.* **1967**, *102*, 533-538.
11. Price, R.G.; Dance, N.; Richards, B.; Cattell, W.R. *Clin. Chim. Acta.* **1970**, *27*, 65-72.
12. Dance, N.; Price, R.G.; Cattell, W.R.; Landsell, J.; Richards, B. *Clin. Chin Acta.* **1970**, *27*, 87-92.
13. Wellwood, J.M.; Ellis, B.G.; Hall, H.; Robinson, D.R.; Thompson, A.E. *Brit. Med. J.* **1973**, *7*, 261-265.
14. Gopalan, V.; VanderJagt, D.J.; Libell, D.P.; Glew, R.H. *J. Biol. Chem.* **1992**, *267*, 9629-9638.
15. Gopalan, V.; Pastuszyn, A.P.; Galey, W.R., Jr.; Glew, R.H. *J. Biol. Chem.* **1992**, *267*, 14027-14032.
16. Daniels, L.B.; and Glew,R.H. *Clin. Chem.* **1982**,*28*, 569-577.
17. Peters, S.P.; Lee, R.E.; Glew, R.H. *Clin. Chim. Acta.* **1975**, *60*, 391-396.
18. Dale, M.P.; Ensley, H.E.; Kern, K.; Sastry, K.A.R.; Byers, L.D. *Biochemistry* **1985**, *24*, 3530-3539.
19. LaMarco, K.L.; Glew, R.H. *Biochem. J.* **1986**, *236*, 469-476.
20. Lai, L.B.; Gopalan, V.; Glew, R.H. *Anal. Biochem.* **1992**, *200*, 365-369.
21. Distler, J.J.; Jourdian, G.W. *Arch. Biochem. Biophys.* **1977**, *178*, 631-643.
22. Llanillo, M.; Perez, N.; Cabezas, J.A. *Int. J. Biochem.* **1977**, *8*, 557-564.
23. Paez de la Cadena, M.; Rodriguez-Berrocal, J.; Cabezas, J.A.; Perez-Gonzalez, N. *Biochemie* **1986**, *68*, 251-260.
24. Glew, R.H.; Peters, S.P.; Christopher, A.R. *Biochim. Biophys. Acta.* **1976**, *422*, 179-199.
25. Legler, G.; Bieberich, E. *Arch. Biochem. Biophys.*, **1988**, *260*, 429-436.
26. Pocsi, I.; Kiss, L. *Biochem. J.* **1988**, *256*, 139-146.
27. DePetro, J.J. (1987) M.S. Thesis, University of Pittsburgh.
28. Ackers, G.K. *Biochemistry*, **1964**, *3*, 723-730.
29. Daniels, L.B. (1983) Ph. D. Thesis, University of Pittsburgh.
30. Moss, D.W. In " The Isoenzymes, " 2nd Edition; Chapman and Hall: New York, NY, 1982.
31. Blake, C.C.F.; Johnson, L.L.N.; Mair, G.A.; North, A.T.C.; Phillips, D.C; Sharma, V.R. *Proc. R. Soc. Lond. Biol. Sci.* **1967**, *167*, 378-388.
32. Capon, B. *Chem. Rev.* **1969**, *69*, 407-498.
33. Capon, B. *Biochimie* **1971**, *53*, 45-149.
34. Hardy, L.W.; Poteete, A.R. *Biochemistry* **1991**, *30*, 9457-9463.
35. Muraki, M.; Harata, K.; Hayashi, Y.; Machida, M.; Jigami, Y. *Biochim. Biophys. Acta.* **1991**, *1079*, 229-237.
36. Legler, G.; Bieberich, E. *Arch. Biochem. Biophys.* **1988b**, *260*, 437-442.
37. Conchie, J.; Gelman, A.L.; Levy, G.A. *Biochem J.* **1967**, *103*, 609-615.
38. Hackert, M.L.; Jacobson, R.A. *J. Soc. Chem. (Chem. Commun.)* **1969**, 11791.

39. Lalegerie, P.; Legler, G.; Yon, J.M. *Biochemie* **1982,** *64,* 977-1000.
40. LaMarco, K.L. (1986) Ph. D. Thesis, University of Pittsburgh.
41. Gopalan, V.; Daniels, L.B.; Glew, R.H.; Claeyssens, M. *Biochem. J.* **1989,** *262,* 541-548.
42. Segel, I.H. In " Enzyme Kinetics "; Wiley Interscience, New York, NY, 1975, 227-272.
43. Gopalan, V.; Glew, R.H.; Libell, D.P.; DePetro, J.J. *J. Biol. Chem.* **1989,** *264,* 15418-15422.
44. Greenzaid, P.; Jencks, W.P. *Biochemistry* **1971,** *10,* 1210-1222.
45. Dale, M.P.; Koffler, W.P.; Chait, I.; Byers, L.D. *Biochemistry* **1986,** *25,* 2522-2529.
46. Cote, G.L.; Tao, R.Y. *Glycoconjugate J.* **1990,** *7,* 145-162.
47. Li, Y-T.; Carter, B.Z.; Rao, B.N.N.; Schweingruber, H.; Li, S-C. *J. Biol. Chem.* **1991,** *266,* 10723-10726.
48. Goodwin, T.W.; Mercer, E.I. In: " Introduction to Plant Biochemistry "; Pergamon Press, Oxford, UK. **1979,** 227-272.
49. Chester, M.A.; Hultberg, B.; Ockerman, P. *Biochim. Biophys. Acta.* **1976,** *429,* 517-526.
50. LaMarco, K.L.; Glew, R.H. *Arch. Biochem. Biophys.* **1985,** *236,* 669-676.
51. Hackney, D.D. In " The Enzymes "; 1990, Vol. 19, 1-36.
52. Fersht, A.F. In " Enzyme Structure and Mechanism "; W. H. Freeman, San Francisco, 1985, 193-220.
53. Kirby, A.J. *Crit. Rev. Biochem. Mol. Biol.* **1987,** *22,* 283-315.
54. Robinson, D.S. In " Food - Biochemistry and Nutritional Value "; Longman Scientific and Technical, Oxford, U.K., 1990, 60-61.
55. Suzuki, K.; Tanaka, H.; Suzuki, K. In " Current Trends in Sphingolipidoses and Allied Disorders"; Volk, B.W., Ed., Plenum, New York, NY, 1976, 99-114.
56. Burton, B.K.; Ben-Youseph, Y.; Nadler, H.L. *Clin. Chim. Acta.* **1978,** *88,* 483-493.
57. Owada, M.; Sakiyama, T.; Kitagawa, T. *Pediat. Res.* **1977,** *11,* 641-646.
58. Dorfman, A.; Matalon, R. In " The Metabolic Basis of Inherited Disease " ; Stanbury, J.B.; Wyngaarden, J.B.; Fredrickson, D.S., Eds., McGraw Hill, New York, NY, 1972, 1218-1272.
59. Weinreb, N.J. *Fed. Proc.* **1976,** *35,* 537.
60. Abrahams, H.E.; Robinson, D. *Biochem J.* **1969,** *111,* 749-755.
61. Tominaga, F.; Oka, K.; Yoshida, H. *J. Biochem.* **1965,** *57,* 717-720.
62. Williamson, D.G.; Collins, D.C.; Layne, D.S.; Conrow, R.B.; Bernstein, S *Biochemistry* **1969,** *8,* 4299-4304.
63. Williamson, D.G.; Layne, D.S.; Collins, D.C. *J. Biol. Chem.* **1972,** *247,* 3286-3288.
64. Mellor, J.D.; Layne, D.S. *J. Biol. Chem.* **1971,** *246,* 4377-4380.
65. Mellor, J.D.; Layne, D.S. *J. Biol. Chem.* **1974,** *249,* 361-365.
66. Rutenburg, M.; Rutenburg, S.H.; Monis, B.; Teague, R.; Seligman, A.M. *J. Histochem. Cytochem.* **1958,** *6,* 122-129.
67. Montgomery, R.D. In "Toxic Constituents of Plant Foodstuffs"; Liener, I.E., Ed., Academic Press, New York, NY, 1980, 265-294.

68.   Reichardt, P.B.; Clausen, T.P. In " Toxicology of Plant and Fungal Compounds "; Keeler, R.F.; Tu, A.T., Eds., Marcel Dekker, Inc., New York, NY, 1991, 313-333.

69.   Mager, J.; Chevion, M.; Glaser, G. In " Toxic Constituents of Plant Foodstuffs " Liener, I.E., Ed., Academic Press, New York, NY, 1980, 265-294.

70.   Haliwell, B.; Gutteridge, J. In " Free Radicals in Biology and Medicine "; Clarendon Press, Oxford, UK, 1989, 329-332.

71.   Wilson, T.H.; Wiseman, G. *J. Physiol.* **1954,** *123,* 116-125.

72.   Gopalan, V. (1991) Ph.D. Thesis, University of New Mexico.

RECEIVED January 25, 1993

Chapter 8

Nutritional Properties of Pyridoxine-β-D-Glucosides

Jesse F. Gregory III

Food Science and Human Nutrition Department, Box 110370, University of Florida, Gainesville, FL 32611−0370

Several vitamin B6 derivatives, especially pyridoxine, exist in plants as ß-D-glucosides. The major form of glycosylated vitamin B6 is pyridoxine-5'-ß-D-glucoside (PNG) , which comprises between 5 and 80% of the total vitamin B6 in various plants. Isotopic studies have indicated that dietary PNG exhibits ~25% bioavailablity in rats and ~58% in humans, relative to simultaneously ingested pyridoxine. The rate limiting phase of PNG utilization in B6 metabolism is the action of mammalian or microbial ß-glucosidases. In view of this incomplete bioavailability, methods of food analysis should permit measurement of both glycosylated and nonglycosylated species.

The generic term vitamin B6 refers to the group of 2-methyl,3-hydroxy,5-hydroxymethyl-pyridines that exhibit the biological activity of pyridoxine. These compounds also exist as 5'-phosphate esters and ß-glucosides, as discussed below. Vitamin B6 functions coenzymically as pyridoxal 5'-phosphate in over 100 enzymic reactions involving the metabolism of amino acids, glycogen, lipids, and neurotransmitters. In addition, vitamin B6 functions in maintaining the flux of substrate to energy metabolism during periods of caloric deprivation through glycogenolysis and amino acid interconversion and catabolism. Therefore, an adequate intake of vitamin B6 is essential for health, growth, cellular homeostasis, and many aspects of physiological function.

ß-Glucosidic conjugates of pyridoxine exist in many plants but are insignificant animal tissues and animal-derived food products. These compounds comprise a significant percentage of the total dietary vitamin B6 intake. Glycosylated vitamin B6 in common foods is largely comprised of pyridoxine-5'-ß-D-glucoside (PN-5'-ß-D-glucoside). Mammalian and, possibly, bacterial ß-glucosidases release free pyridoxine (PN) which is then available for metabolic utilization. The extent of nutritional bioavailability of vitamin B6 from these conjugated forms depends directly on the extent of enzymatic hydrolysis. In addition, recent findings indicate that the fraction of PN-5'-ß-D-glucoside that does not undergo hydrolysis can influence vitamin B6 metabolism by competitively inhibiting the cellular uptake and overall utilization of other nonglycosylated forms of the vitamin. (Abbreviations: pyridoxine-5'-ß-D-

0097−6156/93/0533−0113$06.00/0
© 1993 American Chemical Society

glucoside, PN-5'-ß-D-glucoside; pyridoxine, PN; pyridoxal, PL; pyridoxamine, PM; pyridoxal 5'-phosphate, PLP; pyridoxamine 5'-phosphate, PMP; 4-pyridoxic acid, 4PA; urindinediphospho-glucose, UDP-glucose.)

The objective of this chapter is to assess the nutritional properties of vitamin B6 glucosides with particular attention to: (a) chemical identity and occurrence in foods; (b) analytical methods; and (c) nutritional and metabolic properties of glycosylated forms of the vitamin. These have been addressed, in part, in previous reviews concerning vitamin B6 analysis (1) and its nutritional bioavailability (2,3).

Occurrence of Vitamin B6-ß-D-Glucosides

Evidence of an unidentified "bound" form of the vitamin was reported over 40 years ago by Siegel et al. (4) and Rabinowitz and Snell (5). In studies of the preparation of foods for microbiological assay of total vitamin B6, Rabinowitz and Snell found that the acid concentration and heating time needed to obtain full assay response were much more severe for plant products than needed for samples of animal origin. The severity of extraction/hydrolysis required for plant-derived food samples corresponded to conditions that are now recognized as necessary for hydrolysis of ß-glycosidic conjugates, rather than hydrolysis of phosphorylated or protein-bound (i.e. Schiff base) forms of the vitamin. These studies served as the basis for the eventual selection of conditions adopted by the Association of Official Analytical Chemists for standardized acid extraction and hydrolysis in the determination of total vitamin B6 in plant-derived foods (6).

Yasumoto et al. (7) reported the existence of "bound" forms of vitamin B6 in various cereal grains and legumes. ß-Glucosidase treatment of aqueous buffer extracts caused the release of an unidentified form(s) of vitamin B6 that was active in the microbiological assay using Saccharomyces uvarum. Pyridoxine was the main form of the vitamin released by ß-glucosidase treatment of rice bran. Fractionation of a rice bran extract by size exclusion chromatography prior to hydrolytic treatment using acid or enzymes yielded further evidence of the presence of a ß-glucosidic conjugate of pyridoxine. Evidence of "bound" vitamin B6 in orange juice was found in similar studies by Nelson et al. (8). Although bound species of PL and PM were found, bound PN comprised the majority of the conjugated forms detected. The unidentified bound form(s) constituted over 70% of the total vitamin B6 in orange juice. Membrane filtration and size exclusion chromatography indicated a molecular weight of the conjugate(s) of less than 3000.

Kabir et al. (9) observed variable quantities of vitamin B6 that became active in differential microbiological assays conducted before and after treatment with almond ß-glucosidase. They used this as the basis for an assay for "glycosylated vitamin B6" in foods (discussed later).

Isolation and identification of a conjugated form of vitamin B6 in rice bran was accomplished by Yasumoto et al. (7). On the basis of UV absorption and proton NMR spectra and chromatographic analysis of hydrolytic products, they identified the compound as 5'-O-(ß-D-glucopyranosyl)-pyridoxine (PN-glucoside;

Figure 1). The conjugate reacted with boric acid, which indicated the availability of the 3-hydroxyl and 4'-hydroxymethyl groups, while it did not react with 2,6-dichloroquinone chloroimide, a reagent specific for phenolic hydroxyls, in the presence of boric acid. These findings confirm the 5'-hydroxymethyl group as the site of glycosylation. Chemically synthesized PN-5'-ß-D-glucoside exhibited spectral, chemical and chromatographic characteristics identical to those of the conjugate isolated from rice bran. Gregory and Ink (*10*) observed this same structure for the vitamin B6 conjugate isolated from alfalfa sprouts.

PN-5'-ß-D-glucoside is the major glycosylated form of naturally occurring vitamin B6 in most plant-derived foods, and it comprises a major portion of total vitamin B6 in many items (Table I; *9-11*). Many studies have been conducted concerning the formation of glycosylated species of PN in germinating seeds and immature seedlings (*10,12-15*). Because many of these studies were conducted using germination in the presence of exogenous pyridoxine added at unphysiologically high concentrations (1-10 mM), the distribution of vitamin B6 compounds detected may not reflect the pattern of naturally occurring forms of the vitamin. Such studies have been useful, however, because they facilitated the detection, isolation, and identification of previously unknown glycosylated species.

Tadera et al. (*12,16*) identified PN-5'-ß-D-glucoside as the major metabolite of PN in both rice and pea seedlings germinated in the presence of 1 mM PN. They also isolated and identified 5'-O-[6-O-(3-hydroxyl-3-methyl-4-carboxybutanoyl)-ß-D-glucopyranosyl]pyridoxine, also termed 5'-O-[(3-hydroxy-3-methyl-glutaryl)-ß-D-glucopyranosyl]pyridoxine (HMG-PN-glucoside), as a minor metabolite in seedlings, although no quantitative data were provided. In further studies involving germination of pea seedlings in the presence of 1 mM PN, Tadera et al. (*17*) isolated another minor PN metabolite and identified it as 5'-O-(6-O-malonyl-ß-D-glucopyranosyl)pyridoxine (malonyl-PN-glucoside). The structural formulas of these compounds are shown in Figure 2. The relative molar proportions of the various forms of vitamin B6 in pea seedlings germinated in 1 mM PN were: PN (1), PN-5'-ß-D-glucoside (30), HMG-PN-ß-D-glucoside (7), malonyl-PN-ß-D-glucoside (1) (*17*). Whether the HMG and malonyl derivatives of PN-5'-ß-D-glucoside exist naturally (i.e. in the absence of exogenously added PN) has not been determined.

Evidence of another form of glycosylated vitamin B6 was reported by Tadera et al. (*18*), who detected a compound (or group of compounds) that yields a response in microbiological assays only after alkaline treatment (0.7 M KOH, 3 hr) followed by pH adjustment and ß-glucosidase treatment. The difference between responses obtained with enzymatic treatment alone and from sequential KOH and enzymatic treatment was attributed to an unidentified conjugate termed "B6X." HPLC analysis of a rice bran extract subjected to this hydrolytic treatment revealed that PN was the form of vitamin B6 involved in this conjugated species. It was suggested that the glucosyl moiety of the conjugate was esterified to an unidentified organic acid. Quantitative analysis by differential microbiological assay indicated that B6X fraction constituted the following percentages of total vitamin B6 in selected foods: cauliflower (0%), spinach (0%), pumpkin (0%), immature broad beans (0%), immature podded

Table I.   Concentration of Pyridoxine-5'-β-D-Glucoside (determined by HPLC) or Glycosylated Vitamin B6 Determined by Microbiological Assay in Selected Foods

Food*	total vitamin B6	PN-glucoside	glycosylated Vitamin B6	% glycosylated
	(nmol/g)	(nmol/g)	(nmol/g)	(%)
HPLC Analysis				
Raw broccoli[1]	16.0	5.6	-------	35.1
Bananas[1]	21.6	1.2	-------	5.5
Peanut butter[1]	12.4	7.15	-------	57.6
Carrots, raw[1]	26.1	18.3	-------	70.1
Green beans, raw[1]	15.9	9.3	-------	58.5
Orange juice, frozen[1]	8.7	6.0	-------	69.1
Pork loin, raw[1]	31.1	0.0	-------	0.0
Cow's milk[1]	3.0	0.0	-------	0.0
Human milk[1]	0.99	0.0	-------	0.0
Potatoes, fresh[2]	5.86	0.91	-------	15.5
Potatoes, 9 mo storage[2]	8.81	4.32	-------	49.0
Microbiological Analysis				
Broccoli, raw[3]	9.94		ND	ND
Broccoli, frozen[3]	7.04		4.62	66
Cauliflower, raw[3]	9.23		0.53	6
Cauliflower frozen[3]	4.97		4.08	82
Green beans, raw[3]	3.55		0.36	10
Green beans, canned[3]	1.66		0.47	28
Carrots, raw[3]	10.1		5.15	51
Orange juice, fresh[3]	2.54		0.95	37
Peaches, canned[3]	0.53		0.12	23
Corn frozen[3]	5.21		0.36	7
White rice, cooked[3]	8.17		1.12	14
Navy beans, cooked[3]	22.5		9.41	42
Peanut butter[3]	17.9		3.2	18

*References: [1]Gregory and Ink (10); [2]Addo and Augustin (11); [3]Kabir et al. (43). "Glycosylated vitamin B6" determined by microbiological assay refers to the response yielded by treatment with −β-glucosidase. ND, not detected.

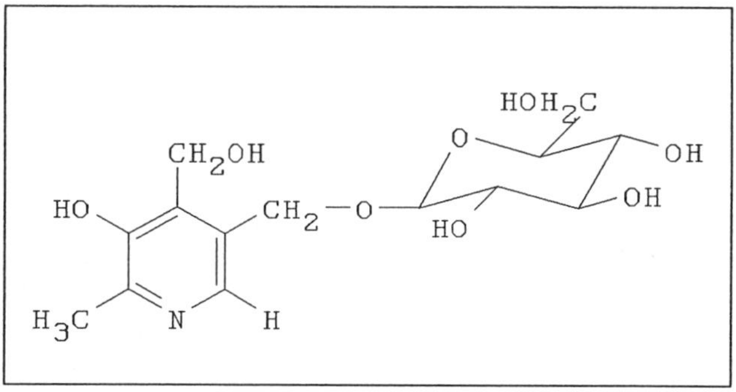

Figure 1. Structural formula of 5'-O-(ß-D-glucopyranosyl)-pyridoxine, the primary form of glycosylated vitamin B6 in foods.

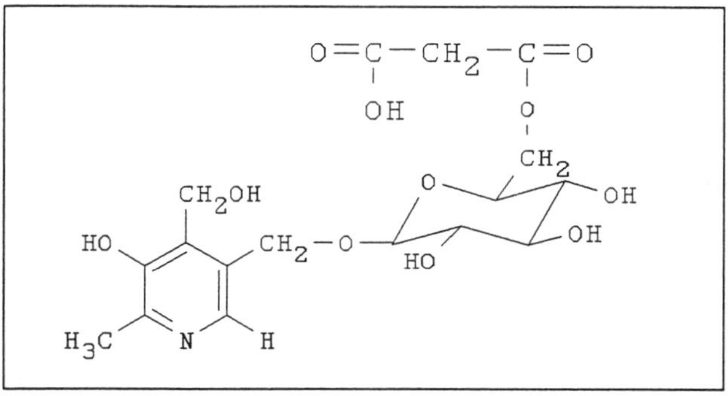

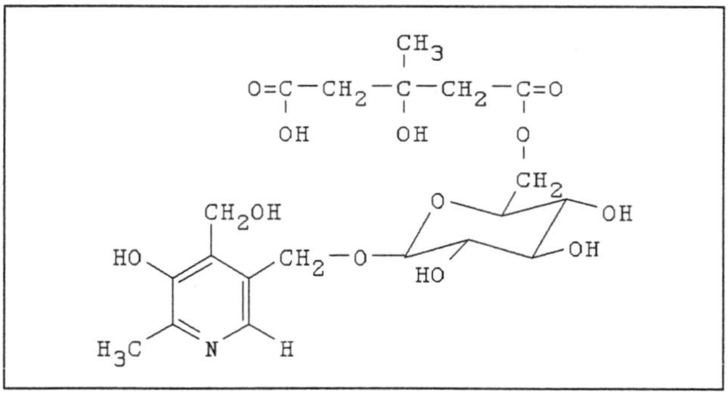

Figure 2. Structural formula of 6-malonyl and 6-(3-hydroxy-3-methyl-glutaryl) esters of PN-5'-ß-D-glucoside. (Adapted from refs. 16 and 17)

peas (21%), defatted soybeans (27%), rice bran (38%), and wheat bran (19%). Tadera and Orite (*19*) isolated a previously unidentified vitamin B6 conjugate from rice bran which exhibited the properties of B6X. They identified it on the basis of spectral, chromatographic, and chemical analyses as 5'-O-[6-O-((+)-5-hydroxy-dioxindole-3-acetyl)ß-cellobiosyl]pyridoxine (Figure 3). Presumably, this derivative, along with the HMG and malonyl esters of PN-5'-ß-D-glucoside (Figure 2), would constitute the B6X fraction in view of the susceptibility to alkaline hydrolysis of their ester linkages, which would render them available to attack by ß-glucosidase.

The molecular site of glycosylation of PN in plants has been examined in studies by Suzuki et al. (*13-15*). These researchers examined 24 plant species cultured in the presence of 10 mM PN and observed both 4' and 5' isomers of PN-ß-D-glucoside. PN-5'-ß-D-glucoside was the major conjugate in most of the species examined, but the quantities of 4' and 5' glucosides were equivalent in wheat and barley (*13,15*). In vitro studies of cultured soybean callus and somatic cells cultured in 10 mM PN yielded PN-5'-ß-D-glucoside as the sole isomer, while rice callus yielded equal amounts of 4' and 5' isomers as seen in intact cereal grain plants discussed above (*14*). No evidence of naturally occurring PN-4'-ß-D-glucoside in plant-derived foods has been reported.

The germination of alfalfa seeds in the presence of added PN serves as a convenient method for the biological synthesis of PN-5'-ß-D-glucoside. Gregory and Ink (*10*) observed that PN-5'-ß-D-glucoside comprised approximately 70-80% of the total naturally occurring vitamin B6 in alfalfa sprouts. A similar percentage of was observed when germination was conducted in the presence of micromolar concentrations of [^{3}H]pyridoxine, and effective glycosylation was also obtained when approximately 50 mg of PN is added per 10 g of dry alfalfa seeds. This biological synthesis, coupled with purification using preparative reverse phase high performance liquid chromatography (HPLC; *10*) or conventional ion-exchange and gel filtration chromatography (*20*) permits convenient preparation of PN-5'-ß-D-glucoside in unlabeled, tritiated, or deuterium labeled form, depending on the form of PN added. When micromolar concentrations of [^{3}H]pyridoxine are used, the specific radioactivity of the PN-5'-ß-D-glucoside produced is considerably lower than the starting material due to isotopic dilution from de novo vitamin B6 synthesis by the plant. Isotopic dilution is negligible when using higher concentrations of PN, as in the synthesis of deuterium-labeled PN-5'-ß-D-glucoside (*21*).

Three additional PN-glycosides have been reported in rice bran: (Figure 4; *22*). Structural identification was based on spectral, chemical, chromatographic, and enzymatic analysis. The significance of these compounds as components of the total vitamin B6 in foods other than rice bran is presently unclear. As stated previously, PN-5'-ß-D-glucoside is the major glycosylated form of vitamin B6 detected in chromatographic studies of a variety of foods to date (*10,23*).

The glycosylation of vitamin B6 also has been found to involve other sugars and different anomeric forms, although there appears to be little metabolic/nutritional significance of these compounds. Incubation of the vitamin B6 catabolite, 4-pyridoxic acid, with homogenates of rat liver has been shown to yield 4-pyridoxic acid-5'-α-D-glucoside (*24*). Similarly, incubation of a rat liver

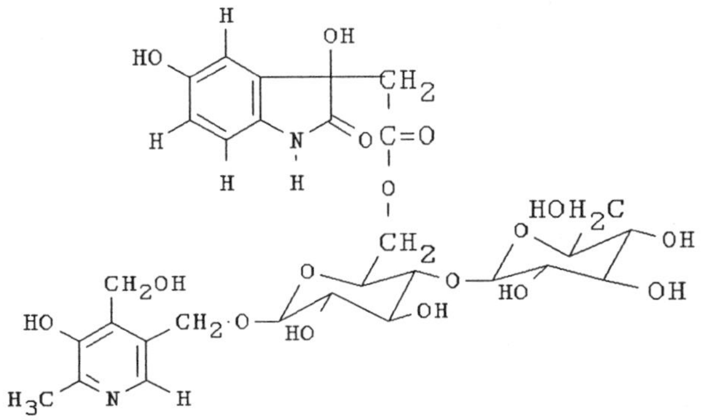

Figure 3. Structural formula of 5'-O-[6-O-((+)-5-hydroxy-dioxindole-3-acetyl)-ß-cellobiosyl]pyridoxine.   This is the major **B6X** fraction in rice bran.   (Adapted from ref. 19)

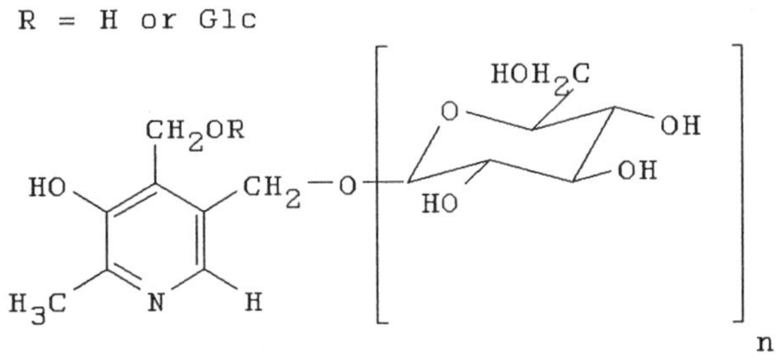

Figure 4.   Structural formulas of pyridoxine oligosaccharides.   Forms identified include 5'-O-(ß-cellobiosyl)-pyridoxine, 4'-O-(ß-D-glucosyl)-5'-O-(ß-cellobiosyl)-pyridoxine,   and   5'-O-(ß-cellotriosyl)-pyridoxine. (Adapted from ref. 22).

homogenate with pyridoxic acid and maltose yielded the 5'-α-maltoside derivative of 4-pyridoxic acid (25). Neither of these compounds is known to be formed in vivo, however. Transglucosylation in bacterial cultures of Sarcina lutea grown in media comprised of PN and sucrose was reported to yield PN-α-D-glucoside (26), but this unusual process has not been detected in bacteria of relevance to foods or the mammalian gastrointestinal tract. In a similar process, the yeast Sporobolomyces singularis produces several β-galactosyl conjugates of PN when cultured in the presence of lactose and PN, and two molds (Aspergillus niger and A. sydowi) produce fructosylpyridoxines when cultured in the presence of sucrose and PN (27). While it is conceivable that such processes may produce vitamin B6 derivatives of occasional nutritional significance, the various β-glucosides discussed previously are clearly the most significant with respect to human or animal nutrition.

Enzymatic Formation of Pyridoxine-β-Glucosides

The mechanism of the formation of glycosylated vitamin B6 in plants has not been determined. As demonstrated using the isotopically-labeled preparations described above, PN-5'-β-D-glucoside can be formed in alfalfa by direct glycosylation of PN. In addition, several research groups have shown the ability of enzymes derived from plants to catalyze the formation of PN-β-D-glycosides from PN as the acceptor (i.e. aglycone) and a suitable β-glucoside donor. Suzuki et al. (13,28) demonstrated that β-glucosidase from wheat bran catalyzed the formation of two glycosylated derivatives, identified as 4' and 5' isomers of PN-β-D-glucoside, during in vitro incubation of PN with cellobiose. Quantitative analysis was not reported but the apparent yield of these products was similar. Studies by Iwami and Yasumoto (29) demonstrated that three isoforms of β-glucosidase isolated from rice bran catalyzed glycosylation of PN, also with the production of both 4' and 5' isomers. PN and a variety of other forms of vitamin B6 could serve as glucosyl acceptors in this reaction, and either cellobiose or p-nitrophenyl-β-D-glucoside were effective donors. p-Nitrophenyl-β-D-glucoside exhibited a K_m of 0.4 mM, while cellobiose yielded a K_i of 16 mM as a competitive inhibitor.

An enzyme in the particulate fraction of pea seedlings also has been shown to catalyze transfer reactions to form PN-5'-β-D-glucoside (30,31). This enzyme is concentrated mainly in the particulate fraction that is sedimented between 20,000 and 50,000 x g in preparative centrifugation, and exhibits K_m values of 0.4 mM for PN and 0.7 mM for uridinediphospho (UDP)-glucose (30). Although this enzyme was similar to those from brans as described above by virtue of its ability to glycosylate forms of vitamin B6 other than PN, it differed in its formation of only the 5'-β-glucosyl isomer of PN-β-D-glucoside (31). The particulate enzyme also exhibited remarkably different substrate specificity for the glucosyl donor. In evaluation of various compounds as potential glucosyl donors, including UDP-glucose, ADP-glucose, GDP-glucose, glucose-1-phosphate, methyl-β-D-glucoside, o-nitrophenyl-β-D-glucoside, cellobiose, and glucose, formation of PN-5'-β-D-glucoside only occurred with UDP-glucose. The enzyme also appeared to exhibit some degree of specificity for certain vitamin B6 compounds (pyridoxine and pyridoxamine), but it did not catalyze glucosylation of o-nitrophenol, riboflavin, thiamin, quercetin, or glucose from UDP-glucose. In total, these results indicate that the B6-conjugating enzyme in pea seedlings is predominantly a glucosyltransferase, rather than a ß-glucosidase (*31*).

Measurement of Glycosylated Forms of Vitamin B6

Because glycosylated forms of vitamin B6 comprise an important fraction of the total vitamin B6 in plant-derived foods, their inclusion in the measurement of vitamin B6 in foods is an analytical problem with important nutritional significance. In view of the incomplete bioavailability of the B6 glucosides it is preferable to select an analytical method that provides separate measurement of nonglycosylated and glycosylated forms of the vitamin. The measurement of vitamin B6 in foods can be performed using analytical approaches that mainly involve microbiological growth methods and HPLC techniques (*1*).

Traditional methods of assay for total vitamin B6 are based on acid hydrolysis in dilute HCl at 121°C (*6*). As a modification of this method, Yasumoto et al. (*32*) and Kabir et al. (*9*) developed similar microbiological assay procedures that involve measurement of microbiologically active vitamin B6 before and after treatment of an aqueous extract of the food sample with almond ß-glucosidase. Although this permits a differential assessment of glycosylated forms of the vitamin, it may also underestimate the concentration of glycosylated vitamin B6 present in raw samples that contain endogenous ß-glucosidase activity. For example, Kabir et al. (*9*) reported that the content of glycosylated vitamin B6 in frozen blanched samples (broccoli, cauliflower, and green beans) markedly exceeded that of raw samples. The use of an extractant that inactivates endogenous ß-glucosidase activity, e.g. perchloric or trichloroacetic acid, would circumvent this problem.

The validity of microbiological assays designed to determine the quantity of total and glycosylated vitamin B6 is predicated on a lack of response of the yeast to intact PN-ß-glucoside(s). Andon et al. (*33*) evaluated the response of the commonly used yeast Saccharomyces uvarum in the measurement of glycosylated vitamin B6 using the assay method of Kabir et al. (*9*). This organism yielded little or no response to PN-5'-ß-D-glucoside prior to hydrolysis. In recovery studies involving the addition of purified PN-5'-ß-D-glucoside to homogenized food composites prior to extraction, 90.2 and 95.9 % recovery was obtained when using enzymatic or acid-catalyzed methods of hydrolysis, respectively. These findings support the validity of the differential microbiological method for samples in which there is little endogenous ß-glucosidase activity.

HPLC is well suited to the direct quantification of vitamin B6 in foods (*1*). The various forms of vitamin B6 may be separated using reverse phase or ion exchange HPLC modes, and the native fluorescence of the pyridine ring permits sensitive fluorometric detection. The key to successful HPLC methods for measurement of vitamin B6 is the preparation of the sample to provide complete extraction and either (a) prevent interconversion of the forms present, or (b) alternatively, accomplish the intentional conversion to species that are more readily measured (e.g. intentional hydrolysis of phosphorylated and/or glycosylated species).

Of the various methods developed for HPLC analysis of vitamin B6 in foods, only the reverse phase approach has been adapted to the measurement of glycosylated forms of the vitamin. Following development of a reverse-phase, ion-pair fluorometric HPLC method for the measurement of PL, PN, PM, PLP, and PMP in foods (*34*), the method was modified to incorporate the measurement of PN-5'-ß-D-glucoside (*10*). PN-5'-ß-D-glucoside was measured either by direct analysis in intact form, in which PN-5'ß-D-glucoside eluted between PN and PL, or as the incremental increase in the PN peak following treatment of the sample extract with almond ß-glucosidase (Figure 5; *10*). The use of 4'-deoxypyridoxine as an internal standard improved analytical precision and compensated for dilution associated with enzymatic treatments and sample purification. Direct measurement of intact PN-5'-ß-D-glucoside requires the availability of the purified compound for use as a standard. It was found that PN and PN-5'-ß-D-glucoside exhibited the same molar fluorescence. The disappearance of the PN-5'-ß-D-glucoside peak was the only significant change caused by enzymatic treatment of common fruits and vegetables, including (broccoli, bananas, green beans, orange juice, and alfalfa sprouts (*10*). Tadera and Naka (*35*) have reported an isocratic ion-pair reverse phase HPLC method that also would be suitable for this type of analysis.

The HPLC method of Gregory and Ink (*10*) was modified and extended to permit measurement of total (free, phosphorylated, and glycosylated) vitamin B6 as well as permitting selective determination of the proportions of PN-5'-ß-D-glucoside and other vitamin B6 ß-glucosides in certain cereal grains and legumes (*23*). These studies were conducted because of the demonstrated existence of several forms of glycosylated PN other than PN-5'-ß-D-glucoside (i.e. PN-oligosaccharides), along with the esterified forms of PN-glucosides. The original HPLC procedure was modified because of changes in the properties of the commercial reverse phase HPLC column originally employed (Perkin-Elmer "3x3"), apparently caused by commercial reformulation, that lessened the retention of PLP, the first-eluting peak (*1,23*).

The modified HPLC method (*23*) involved the use of a reverse-phase ion-pair separation using a Beckman Ultrasphere IP column, with analysis of plant-derived foods using the differential approach devised by Tadera et al. (*18*) to accomplish selective hydrolysis of the various forms of the vitamin. This protocol involved: (a) direct analysis of extracts, which permitted direct measurement of PN, PL, PM, PLP, PMP, and PN-5'-ß-D-glucoside; (b) analysis of ß-glucosidase treated extracts, to permit measurement of PN-5'-ß-D-glucoside and other PN-ß-glucosides as the incremental increase in PN; and (c) analysis of extracts following sequential treatment with KOH and ß-glucosidase, to permit measurement of saponifiable PN-glucoside esters such as the malonyl or HMG esters of PN-5'-ß-D-glucoside or the B6X compound (*23*). The results from the analysis of selected food samples are presented in Table II (*23*). Although evidence of PN-oligosaccharides and esters was observed in these samples, common fruits and vegetables contain PN-5'-ß-D-glucoside as the primary glycosylated form of the vitamin.

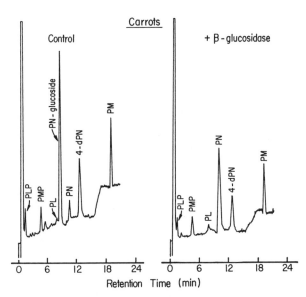

Figure 5. HPLC determination of vitamin B6 in carrots after treatment with ß-glucosidase. (Reproduced with permission from ref. 10. Copyright 1987 American Chemical Society)

Table II. Relative concentration of glycosylated species in selected foods, as determined by HPLC after selective hydrolytic treatments

Sample	Glycosylated	PN-5'-β-Glc	B6X*
	------- (percentage of total vitamin B6) -------		
Soybean flour	67	67	ND
Oat bran	ND	ND	ND
Wheat bran	45	34	11
Rice bran	82	36	43
Carrots, raw	77	65	12

*These data were adapted from ref. 23. B6X refers to esterified glycosylated forms of PN released by sequential treatment with KOH and β-glucosidase. ND, not detected.

Nutritional Properties of Vitamin B6-ß-Glucosides

The in vivo function of the various forms of dietary vitamin B6 depends on the extent of their metabolic conversion to the coenzymic form, pyridoxal 5'-phosphate. As shown in Figure 6, the various forms of vitamin B6 are interconvertible. Following intestinal absorption in nonphosphorylated form (i.e. as PL, PM, or PN), conversion to PLP occurs by phosphorylation followed by oxidation catalyzed by pyridoxal kinase and pyridoxine (pyridoxamine) 5'-phosphate oxidase, respectively (36). For PN-ß-D-glucoside(s) to serve as a source of dietary vitamin B6, absorption and hydrolysis of the glycosidic bond (not necessarily in that order) must occur to release metabolizable PN (Figure 6). Assessment of the nutritional properties of glycosylated vitamin B6 has involved investigation of its absorption and hydrolysis, as well as its influence on other phases of vitamin B6 transport and metabolism. The adequacy of vitamin B6 nutritional status depends on the quantity of dietary vitamin B6 consumed as well as its bioavailability, i.e. the extent of intestinal absorption and metabolic utilization (3). Inadequate vitamin B6 nutrition prevalent among Americans, especially women, is based mainly on the low intake relative to Recommended Dietary Allowance values (37). Incomplete bioavailability would intensify the problem of inadequate intake. Although the overall bioavailability of vitamin B6 in foods has not been totally determined, research evidence suggests that the incomplete utilization of glycosylated vitamin B6 is a major factor affecting the overall bioavailability of dietary vitamin B6. Because PN-5'-ß-D-glucoside comprises 25-75% of the vitamin B6 in many common fruits and vegetables, a thorough understanding of its nutritional properties is important.

The metabolic utilization of glycosylated vitamin B6 requires absorption and hydrolysis of the glycosidic bond. The hydrolysis is an obligatory and probably rate limiting step in the utilization of vitamin B6 glucosides. Intestinal absorption of vitamin B6 occurs by passive diffusion of vitamin B6 compounds in nonphosphorylated form (38), and absorption of intact PN-5'-ß-D-glucoside also appears to be a passive diffusion process (20,29).

The nutritional properties of glycosylated vitamin B6 were first examined by Tsuji et al. (39). Chemically synthesized PN-5'-ß-D-glucoside was found to be efficiently absorbed in vitro by everted rat jejunum, and the appearance of free PN in serosal fluid provided evidence of partial hydrolysis. The in vivo nutritional activity of PN-5'-ß-D-glucoside appeared to be equivalent to free PN with respect to the response of various vitamin B6-dependent enzymes and xanthurenic acid excretion following oral or intravenous administration of PN-5'-ß-D-glucoside. It should be noted that the quantity administered (30 μg PN equivalents) exceeded the vitamin B6 requirement for growing rats. Thus, this experimental design may not have effectively differentiated between complete and only partial utilization of the PN-5'-ß-D-glucoside. Further in vitro studies showed equivalent absorption of PN-4'-ß-D-glucoside and free PN (29). Incubation in the presence of ß-glucosidase inhibitor δ-gluconolactone had no influence on the absorption of PN-4'-ß-D-glucoside, which indicated that hydrolysis per se had no effect on the transmural transport.

A series of in vivo studies with rats orally administered PN and PN-5'-ß-D-

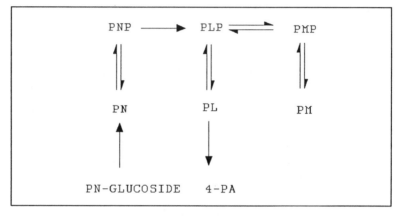

Figure 6. Schematic of metabolic interconversions of various forms of vitamin B6. The hydrolysis of the ß-glycosidic bond must occur before the pyridoxyl moiety of a PN-glucoside can enter vitamin B6 metabolism.

glucoside provided additional information concerning the comparative absorption and metabolism of these compounds. Studies involving simultaneous administration of PN and PN-5'-β-D-glucoside indicated effective and nearly equivalent absorption in vivo (40,41). The in vivo metabolic utilization of the orally administered PN-5'-β-D-glucoside was approximately 20-30% as complete as that of PN, and intact PN-5'-β-D-glucoside that did not undergo in vivo hydrolysis was rapidly excreted in the urine (20,40,41). This limited utilization of PN-5'-β-D-glucoside in vitamin B6 metabolism was substantially lower than observed by Tsuji et al. (39); the reasons for this difference have not been determined. PN-5'-β-D-glucoside administered to rats by intraperitoneal injection underwent less than half of the metabolic utilization observed with oral administration, which indicates a significant role of the intestine in hydrolyzing PN-5'-β-D-glucoside during or prior to absorption (40,41). In spite of limited utilization, however, administration of sufficiently high concentrations of dietary PN-5'-β-D-glucoside to rats can adequately support nutritional requirements for growth (20).

The bioavailability of purified PN-5'-β-D-glucoside in humans was examined using deuterium-labeled forms of PN and PN-5'-β-D-glucoside (21). After concurrent oral administration of [^{2}H$_5$]PN and [^{2}H$_2$]PN-5'-β-D-glucoside mixed in oatmeal, their relative metabolic utilization was evaluated from the molar ratio of labeled forms of urinary 4-pyridoxic acid determined by mass spectrometry. By this technique the utilization of PN-5'-β-D-glucoside was 58 ± 13% relative to oral PN. When PN-5'-β-D-glucoside was administered by intravenous infusion (60 min), its utilization was only 28 ± 6% relative to oral PN. These studies indicated that dietary PN-5'-β-D-glucoside would contribute significantly to our intake of nutritionally available vitamin B6 (i.e. ~58% as effective as free PN).

The bioavailability of PN-5'-β-D-glucoside is incomplete as a result of its incomplete hydrolysis. Controlled dietary studies evaluating total dietary PN-5'-β-D-glucoside and urinary excretion of intact PN-5'-β-D-glucoside by human subjects indicated urinary excretion of approximately 35% of the ingested PN-5'-β-D-glucoside (21). This observation is consistent with the isotopically derived assessment of PN-5'-β-D-glucoside bioavailability.

In addition to the studies of PN-5'-β-D-glucoside bioavailability described above, studies of the overall bioavailability of vitamin B6 in foods have provided insight into the nutritional properties of glycosylated vitamin B6. Tarr et al. (42) examined the utilization of vitamin B6 in a mixed American diet using a long-term metabolic study with human subjects and observed a mean bioavailability of approximately 75% relative to PN in a formula diet. Although the factors responsible for this incomplete bioavailability were not determined, this study was important because it indicated that the net bioavailability of vitamin B6 in a typical diet is incomplete but reasonably high. Kabir et al. (43) used a short-term bioassay with human subjects to examine the bioavailability of vitamin B6 in various individual foods. They reported a negative correlation between the percentage of glycosylated B6 and the apparent bioavailability of the vitamin in canned tuna, whole wheat, and peanut butter. Further studies in the same laboratory (44) with additional foods showed that this correlation was inconsistent in the case of certain items. Factors including the influence of food composition on intestinal transit time may have been involved. The role of glycosylated vitamin B6 in maternal and infant nutrition has been evaluated in rats and humans. Orally administered tritiated PN-5'-ß-D-glucoside undergoes limited metabolic utilization by lactating rats, although intact PN-5'-ß-D-glucoside is not secreted in the milk (45). Little or no PN-5'-ß-D-glucoside is typically present in human milk from American subjects, whether vegetarian or omnivorous (10,33). This is in contrast to the report that glycosylated species comprised 15% of the total vitamin B6 in milk samples from Nepalese vegetarian women (46), a difference that has not been explained. Andon et al. (33) assessed the diet composition and nutritional status of lactating American women and their infants. Glycosylated vitamin B6 comprised approximately 15% of the maternal diets, and, over the range examined, the quantity of glycosylated vitamin B6 had no statistically detected influence on the vitamin B6 nutritional status of women or infants. This finding is consistent with other reports of incomplete bioavailability.

ß-Glucosidase activity is present at low levels in various mammalian tissues (47-49). Although the intestinal ß-glucosidase responsible for in vivo hydrolysis of PN-5'-ß-D-glucoside has not been isolated and characterized, this enzyme has been shown to be highly similar to the cytosolic broad specificity ß-glucosidase in other mammalian tissues (50). The intestinal ß-glucosidase activity capable of hydrolyzing PN-5'-ß-D-glucoside was mainly cytosolic, optimally active at pH 6, and was inhibited by sodium taurocholate. The enzyme exhibited a high Km for PN-5'-ß-D-glucoside ($>2\,\mathrm{mM}$) and substantially lower rates of hydrolysis with PN-5'-ß-D-glucoside than with the synthetic substrate 4-methyl-umbelliferyl-ß-D-glucoside (50). The specific activity of the enzyme was approximately 6-fold greater in human small intestine than in rat and guinea pig (50), and enzyme activity was correlated with the observed in vivo bioavailability of PN-5'-ß-D-glucoside in these species (50; Banks and Gregory, unpublished). It is not currently possible to assess the role of microbial ß-glucosidase activity in the intestinal hydrolysis of dietary PN-5'-ß-D-glucoside or other vitamin B6 glucosides. However, the jejunum, primary site of vitamin B6 absorption (38), is known to have little microbial colonization in humans. Thus, ß-glucosidase activity in the intestinal mucosal cells appears to be responsible for much of the in vivo release of free PN from PN-5'-ß-D-glucoside in humans. ß-Glucosidase activity of intestinal mucosa is not affected by the concentration of dietary PN-5'-ß-D-glucoside in rats, which indicates a lack of diet-related induction (51). As shown by the partial metabolic utilization of intravenously administered PN-5'-ß-D-glucoside in humans, ß-glucosidase activity in other tissues (presumably including liver) is also involved in PN-5'-ß-D-glucoside hydrolysis (21).

Aside from its incomplete bioavailability as a source of vitamin B6, PN-5'-ß-D-glucoside has been found to exert an influence on the metabolism of free (i.e. nonglycosylated) forms of the vitamin. Gilbert and Gregory (52) observed that PN-5'-ß-D-glucoside retarded the metabolic utilization of [^{14}C]PN in a dose-dependent manner when administered simultaneously in rats. Hepatic retention of labeled forms of vitamin B6 and urinary excretion patterns provided evidence of an antagonistic effect of PN-5'-ß-D-glucoside (Figure 7). Similar evidence of

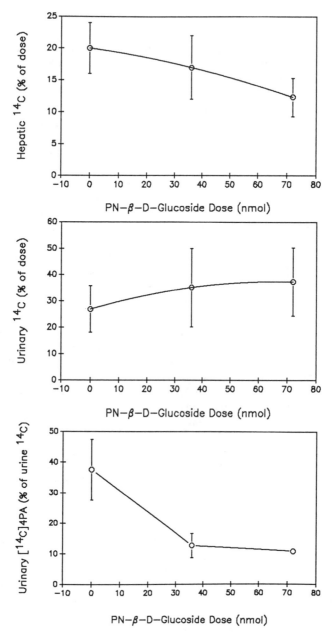

Figure 7. Antagonistic effect of PN-5'-ß-D-glucoside on the retention and metabolism of [¹⁴C]PN in rats. (Adapted from ref. 52)

this antagonism was observed in human subjects in an experiment evaluating the metabolism of deuterium-labeled PN based on urinary excretion of labeled 4-pyridoxic acid (*53*). In vitro studies indicated that PN-5'-ß-D-glucoside is not an inhibitor of PL kinase, pyridoxine (pyridoxamine) 5'-phosphate oxidase, or glycogen phosphorylase b (*54*), which ruled out a direct effect on vitamin B6 metabolism, while lack of inhibition of glycogen phosphorylase b indicated no influence of PN-5'-ß-D-glucoside on this aspect of vitamin B6 coenzymic function. Recent studies have indicated that PN-5'-ß-D-glucoside competitively inhibits the uptake of PN into isolated rat liver cells (*55*). This effect occurs at physiologically relevant concentration (Ki=1.4 μM) and appears to explain the competitive effect of PN-5'-ß-D-glucoside in vivo. This action of PN-5'-ß-D-glucoside is the first antagonism affecting vitamin B6 that does not involve inhibition of PL kinase, but rather occurs at the site of the vitamin B6 transport protein.

SUMMARY AND CONCLUSION

Further clarification of the nutritional significance of PN-5'-ß-D-glucoside and other glycosylated forms of vitamin B6 requires more complete analytical data concerning the forms of vitamin B6 in foods. Further assessment of the factors affecting the extent of utilization of glycosylated species in vitamin B6 metabolism, as well as the importance of the competition between free and glycosylated forms of the vitamin, is also needed to permit a full understanding of the nutritional impact of glycosylated vitamin B6.

ACKNOWLEDGMENTS

The author gratefully acknowledges support of this research by Grant No. 83-CRCR-1-1240 from the Competitive Research Grants Office, U. S. Department of Agriculture, and Grant No. DK-37481 from the National Institutes of Health.

LITERATURE CITED

1. Gregory, J. F. *J. Food Comp. Anal. 1988*, **1988**, *1*, 105-123.
2. Leklem, J. E. *Food Technol.* **1988**, *42*, 194-196.
3. Gregory, J.F. *Ann. NY Acad. Sci.* **1990**, *585*, 86-95.
4. Siegel, L.; Melnick, D.; Oser, B.L. *J. Biol. Chem.* **1943**, *149*, 361-367.
5. Rabinowitz, J.C.; Snell, E.E. *Ind. Eng. Chem. Anal. Ed.* **1947**, *19*, 277-280.
6. Toepfer, E.W.; Polansky, M.M. *J. Assoc. Offic. Anal. Chem.* **1970**, *53*, 546-550.
7. Yasumoto, K.; Tsuji, H.; Iwami, K.; Mitsuda, H. *Agric. Biol. Chem.* **1977**, *41*, 1061-1067.
8. Nelson, E. W.; Burgin, C.W.; Cerda, J.J. *J. Nutr.* **1977**, *107*, 2128-2134.
9. Kabir, H.; Leklem, J.E.; Miller, L.T. *J. Food Sci.* **1983**, *48*, 1422-1425.
10. Gregory, J. F.; Ink, S.L. *J. Agric. Food Chem.* **1987**, *35*, 76-82.
11. Addo, C.; Augustin, J. *J. Food Sci.* **1988**, *53*, 749-752
12. Tadera, K.; Nakamura, M.; Kobayashi, A. *Vitamins* **1978**, *52*, 17-22.

13. Suzuki, Y.; Inada, Y.; Uchida, K. *Phytochemistry* **1986**, *25*, 2049-2051.
14. Suzuki, Y.; Ishii, H.; Suga, K.; Uchida, K. *Phytochemistry* **1986**, *25*, 1331-1332.
15. Suzuki, Y. *J. Japanese Soc. Starch Sci.* **1988**, *35*, 93-102.
16. Tadera, K.; Nagano, K.; Yagi, F.; Kobayashi, A.; Imada, K.; Tanaka, S. *Agric. Biol. Chem.* **1983**, *47*, 1357-1359.
17. Tadera, K.; Mori, E.; Yagi, F.; Kobayashi, A.; Imada, K.; Imabeppu, M. *J. Nutr. Sci. Vitaminol.* **1985**, *31*, 403-408.
18. Tadera, K.; Kaneko, T.; Yagi, F. *Agric. Biol. Chem.* **1986**, *50*, 2933-2934.
19. Tadera, K.; Orite, K. *J. Food Sci.* **1991**, *56*, 268-269.
20. Trumbo, P.R.; Gregory, J.F.; Sartain, D.B. *J. Nutr.* **1988**, *118*, 170-175.
21. Gregory, J.F.; Trumbo, P.R.; Bailey, L.B.; Toth, J.P.; Baumgartner, T.G.; Cerda, J.J. *J. Nutr.* **1991**, *121*, 177-186.
22. Tadera, K.; Kaneko, T.; Yagi, F. *J. Nutr. Sci. Vitaminol.* **1988**, *34*, 167-177.
23. Gregory, J.F.; Sartain, D.B. *J. Agric. Food Chem.* **1991**, *39*, 899-905.
24. Tadera, K.; Yasumoto, K.; Izawa, Y.; Waki, H.; Mitsuda, H. *J. Nutr. Sci. Vitaminol.* **1973**, *19*, 145-150.
25. Tadera, K.; Yasumoto, K.; Harada, T.; Mitsuda, H. *J. Nutr. Sci. Vitaminol.* **1973**, *19*, 145-150.
26. Ogata, K.; Tani, Y.; Uchida, Y.; Tochikura, T. *Biochim. Biophys. Acta* **1968**, *165*, 578-579.
27. Suzuki, Y.; Uchida, K. *Biochim. Biophys. Acta* **1992**, *1116*, 67-71.
28. Suzuki, Y.; Uchida, K.; Tsuboi, A. *Nippon Nogeikagaku Kaishi* **1979**, *53*, 189-196.
29. Iwami, K.; Yasumoto, K. *Nutr. Res.* **1986**, *6*, 407-414.
30. Tadera, K.; Nakamura, M.; Yagi, F.; Kobayashi, A. *J. Nutr. Sci. Vitaminol.* **1979**, *25*, 347-350.
31. Tadera, K.; Yagi, F.; Kobayashi, A. *J. Nutr. Sci. Vitaminol.* **1982**, *28*, 359-366.

32. Yasumoto, K.; Iwami, K.; Tsuji, H.; Okada, J.; Mitsuda, H. *Bitamin* **1976**, *50*, 327-333.
33. Andon, M.B.; Reynolds, R.D.; Moser-Veillon, P.B.; Howard, M.P. *Am. J. Clin. Nutr.* **1989**, *50*, 1050-1058.
34. Gregory, J.F.; Feldstein, D. *J. Agric. Food Chem.* **1985**, *33*, 359-363.
35. Tadera, K.; Naka, Y. *Agric. Biol. Chem.* **1991**, *55*, 563-564.
36. Merrill, A.H.; Henderson, J.M. *Ann. N.Y. Acad. Sci.* **1990**, *585*, 110-117.
37. Kant, A.K.; Block, G. *Am. J. Clin. Nutr.* **1990**, *52*, 707-716.
38. Henderson, L. M. In: *Vitamin B6: Its Role in Health and Disease*; Reynolds, R.D.; Leklem, J.E., Eds.; Current Topics in Nutrition and Disease. Publisher: Alan R. Liss, Inc., New York, NY, 1985, Vol. 13; pp 25-33.
39. Tsuji, H.; Okada, J.; Iwami, K.; Yasumoto, K.; Mitsuda, H. *Vitamins* **1977**, *51*, 153-159.
40. Ink, S. L.; Gregory, J.F.; Sartain, D.B. *J. Agric. Food Chem.* **1986**, *34*, 857-862.
41. Trumbo, P.R.; Gregory, J.F. *J. Nutr.* **1988**, *118*, 1336-1342.
42. Tarr, J. B.; Tamura, T.; Stokstad, E.L.R. *Am. J. Clin. Nutr.* **1981**, *34*, 1328-1337.

43. Kabir, H.; Leklem, J.E.; Miller, L.T. *Nutr. Rept. Int.* **1983**, *28*, 709-715.
44. Bills, N. D.; Leklem, J.E.; Miller, L.T. *Fed. Proc.* **1987**, *46*, 1487.
45. Trumbo, P.R.; Gregory, J.F. *J. Nutr.* **1989**, *119*, 36-39.
46. Reynolds, R.D. *Am. J. Clin. Nutr.* **1988**, *48*, 863-867.
47. Mellor, J.D.; Layne, D.S. *J. Biol. Chem.* **1974**, *249*, 361-365.
48. Daniels, L.B.; Coyle, P.J.; Chiao, Y.-B.; Glew, R.H.; Labow, R.S. *J. Biol. Chem.* **1981**, *256*, 13004-13013.
49. LaMarco, K.L.; Glew, R.H. *Biochem. J.* **1986**, *237*, 469-476.
50. Trumbo, P.R.; Banks, M.A.; Gregory, J.F. *Proc. Soc. Expt. Biol. Med.* **1990**, *195*, 240-246.
51. Banks, M.A.; Gregory, J.F. *Nutr. Res.* **1991**, *11*, 169-175.
52. Gilbert, J.A.; Gregory, J.F. **1992**. *J. Nutr.* *122*, 1029-1035.
53. Gilbert, J.A.; Gregory, J.F.; Bailey, L.B.; Toth, J.P.; Cerda, J.J. *FASEB J.* **1991**, *5*, A586.
54. Gilbert, J.A.; Gregory, J.F. *FASEB J.* **1992**, *6*, A1373.
55. Zhang, Z.; Gregory, J.F.; McCormick, D.B. *J. Nutr.* **1993** (in press).

RECEIVED January 12, 1993

Chapter 9

β-Glucosidase (Linamarase) of the Larvae of the Moth *Zygaena trifolii* and Its Inhibition by Some Alkaline Earth Metal Ions

Adolf Nahrstedt and Elisabeth Mueller

Institut für Pharmazeutische Biologie und Phytochemie, Westf., Wilhelms-Universität, D-4400 Münster, Germany

ß-Glucosidases involved in the metabolism of endogeneous secondary constituents have rarely been found in insects. One example is the ß-glucosidase (linamarase) of the larvae of *Zygaena trifolii* (Zygaenidae, Lepidoptera) which is a glycoprotein and co-occurs with its substrates, the cyanogenic glucosides linamarin and lotaustralin, in the hemolymph of the insect. *In vitro* experiments showed a noncompetitive or mixed type inhibition of the linamarase by the alkaline earth metal ions Mg^{++} and Ca^{++}. When fully activated by chelating agents the linamarase causes strong cyanogenesis (liberation of HCN) in the hemolymph. Lowering the pH from its physiological value of 6.2 in a non-chelating buffer to 3.6 causes also almost full cyanogenesis. Mg^{++} and Ca^{++} ions occur at 18 and 7 mM, respectively, in the hemolymph; the inhibitor constants (K_I) are 5 mM for Mg^{++} and 80 mM for Ca^{++}. The data suggest that both ions are natural inhibitors of the linamarase in the intact insect.

Although ß-glucosidases capable to catalyze the hydrolyzation of glycosidic secondary metabolites to aglykones and glucose are well known from many plants (*1-3*) and microorganisms (*4*) their occurrence and characterization in insects are rarely documented (*5*). Most of such ß-glucosidases have been obtained from the intestinal tract of several insects; they possess *in vitro* activities towards cellobiose (indicating cellulose degrading activity), or towards salicin, helicin, amygdalin and arbutin, which are usually used to demonstrate ß-glucosidase activity; but they are seldom the natural substrates originating from the food, by sequestration or by *de novo*-synthesis within the insect (*5-8*). Especially xylophagous insects have ß-glucosidases that hydrolyze cellulose (cellobiose) (*5,9*) for energy supply. Subspecies of *Papilio glaucus* (Lepidoptera) adapted to feed on plants rich in phenolics contain higher activity of 1,4-ß-glucosidase to degrade salicin and salicortin more efficiently than do

0097–6156/93/0533–0132$06.00/0

non-adapted populations (*10*). Other insects are able to select their food plant in that they hydrolyze glycosidic allelochemicals of their host as it was observed for the peachtree borers (*11*) which metabolize the cyanogenic glucoside prunasin; this activity was also inducible during feeding on an amygdalin containing diet (*11*). Such examples demonstrate the advantage of endogeneous ß-glucosidases for adapted insects making them able to metabolize exogeneous more or less toxic metabolites for their own benefit. For the "normal" non-adapted feeder, however, numerous plant glycosides possess antifeedant, oviposition deterrent or toxic activities towards herbivorous insects produced by their corresponding aglykones as has been observed for glucosinolates (*12*), cyanogenic glycosides (*13*) or cardiac glycosides (*14*). It is, however, not clear how far exogeneous ß-glucosidases of the host or endogeneous ß-glucosidases of the herbivore contribute to the liberation of the respective aglykone.

Only a few reports deal with ß-glucosidases involved in the metabolism of endogeneous secondary metabolites of insects. Cockroaches (*Blatta orientalis, Periplaneta americana*) produce a milky secretion in their left colleterial gland that contains the 4-O-glucoside of 3,4-dihydroxybenzoic acid, a polyphenol oxidase and a structural protein. The right gland, however, produces a clear secretion containing a ß-glucosidase activity. When both secretions are mixed free protocatechuic (3,4-dihydroxybenzoic) acid is released from its glucoside and is used for the production of sclerotin after oxidation and reaction with the supplied protein to form the ootheca (*15*). Some members of the Diptera (*16,17*) and Lepidoptera (*18-21*) produce ß-D-glucopyranosyl-O-L-tyrosine that is primarily synthesized during the late larval feeding period and accumulates in highest levels prior to ecdysis; tyrosine is released from its glucoside and used to produce quinones and catechols to tan the new cuticle; in *Manduca sexta* hydrolysis is catalyzed by a ß-glucosidase, that is preferentially localized in the fat body and is under the control of 20-hydroxyecdysone (*22*).

Cyanogenesis in Zygaena Species

All hitherto investigated species of the lepidopteran moth family Zygaenidae, several species of the butterfly family Nymphalidae and some of the Lycaenidae are cyanogenic (for review see *23* and *24*). With the exception of *Acraea horta* (Nymphalidae) (*25*) all species investigated so far contain the cyanogenic glucosides linamarin and lotaustralin or linamarin alone (Figure 1). One of the best investigated species of the cyanogenic Lepidoptera is *Zygaena trifolii* (Esper 1783) whose larvae were used for many experiments. It was shown that linamarin and lotaustralin are not only synthesized *de novo* by the larvae starting from valine and isoleucine as biogenetic precursors (*24*) in close similarity to the pathway used by higher plants (*26*), but it was also demonstrated that the larvae are able to sequester both glucosides from their host, *Lotus corniculatus* (Fabaceae), which also accumulates linamarin and lotaustralin (*27*).

When the larvae are injured hydrogen cyanide is liberated indicating the activity of degrading enzymes such as ß-glucosidases and, eventually, hydroxynitrile lyases (HNLs) as is well known from cyanogenic higher plants (*3*): When hydrolyzed by a ß-glucosidase the substrates linamarin and lotaustralin are converted to the corresponding cyanohydrins (hydroxynitriles) acetone cyanohydrin and methylethyl

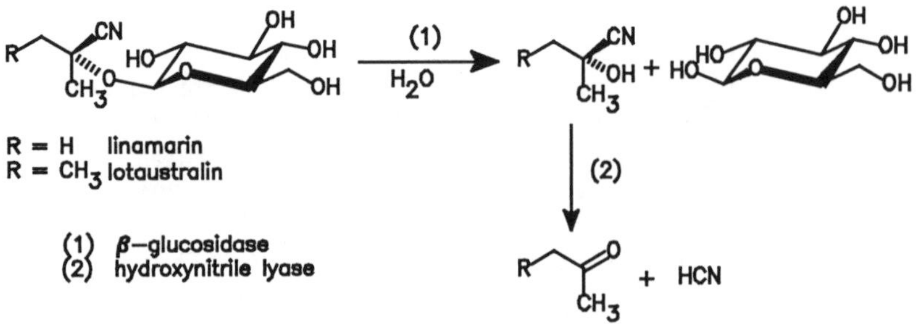

R = H   linamarin
R = CH₃ lotaustralin

(1) β−glucosidase
(2) hydroxynitrile lyase

Figure 1. Degradation of linamarin and lotaustralin catalyzed by a ß-glucosidase and a hydroxynitrile lyase, so called cyanogenesis.

ketone cyanohydrin; both decompose spontaneously at pH values above ca. 5.5 to give acetone, methylethyl ketone and hydrogen cyanide (Figure 1). Thus, for cyanogenesis a ß-glucosidase is sufficient; HNLs, however, accelerate the final step of cyanogenesis up to 20-fold as observed for the HNL of *Hevea brasiliensis* (Euphorbiaceae) (*28*) and the HNL of *Zygaena trifolii* (*29*).

When testing the different tissues of a dissected larva of *Z. trifolii*, using the endogenous substrate linamarin, it became evident that both enzyme activities are present preferentially in the hemolymph whereas other organs contain small or negligible amounts of activity (Table 1). Thus, the larvae possess the entire cyanogenic system as do cyanogenic higher plants. Both enzymes, the ß-glucosidase (*30*) and the hydroxynitrile lyase (*29*, Mueller and Nahrstedt in preparation) have been isolated from the hemolymph, purified and characterized.

Table I. Distribution of Cyanoglyucosides, ß-Cyanoalanine (ß-CNA), Activity of Linamarase, Hydroxynitrile Lyase (HNL) and ß-Cyanoalanine Synthase (ß-CAS) in a Larval Body of *Zygaena trifolii*

Organ	Cyanogluco-sides	Linamarase [activity]	HNL [activity]	ß-CAS [activity]	ß-CNA
Hemolymph	33.3%	84%	82%	<1%	74%
Integument	65.8%	11%	3.7%	22%	16%
Secretion	19.6%	0%	0%	<1%	0.36%
Fat body	0.25%	2%	8.8%	27%	5.5%
Gut epithel	<0.1%	0.5%	1.8%	12%	0.94%
Gut content	n.d.	2%	3.5%	39%	2.47%
Malpighii	<0.1%	0%	0.3%	<1%	0.33%
Silk gland	0.13%	0%	0%	n.d.	n.d.

The Linamarase of the Larvae of *Zygaena trifolii*

The ß-glucosidase is a dimer of ca. 130 KDa. It shares some properties with plant ß-glucosidases that hydrolyze linamarin and lotaustralin (*31-34*): i. The molecular weight of its subunits is ca. 66 kDa and is in the order of those of higher plants (ca. 60 - 70 kDa); ii. its K_M value for linamarin is 7.8 mM (plants between 4.3 and 7.6 mM); iii. the pH optimum is at 4.5 to 5 (plants from 5-6).

Many plant ß-glucosidases hydrolyzing linamarin and lotaustralin, however, exhibit a broad substrate specificity; the *Zygaena* enzyme, in contrast, is selective (Table 2) for ß-monoglucosides with an aliphatic cyanogenic side chain but does not or only weakly catalyze the hydrolysis of i. cyanogenic monoglucosides with a cyclic (acalyphin) or aromatic (prunasin) aglykone and non-cyanogenic aromatic monoglucosides such as salicin; ii. cyanogenic aromatic (amygdalin) and aliphatic (linustatin, the glucoside of linamarin) diglucosides (gentiobiosides); iii. exogeneous (cellobiose) or endogeneous (trehalose) carbohydrates used for energy supply.

Table II. Specificity of the Linamarase obtained from the Hemolymph of the Larvae of *Zygaena trifolii*

Substrate	K_M mmol/L	V_{max} U/mg	Relative Activity
linamarin	7.8	71.0	50 %
lotaustralin	2.5	65.0	105 %
prunasin	0.0	21.2	7.1 %
acalyphin			<< 1 %
4-NP-ß-D-glucoside	0.27	20.9	100 %
4-NP-α-D-glucoside	-	-	<< 1 %
4-NP-ß-D-galactoside	-	-	0 %
salicin	-	-	<< 1 %
trehalose	-	-	0 %
cellobiose	-	-	0 %

In summary the enzyme is adapted to the hydrolysis of the endogeneous substrates linamarin and lotaustralin to give finally hydrogen cyanide and is thus correctly called a "linamarase".

Although a turnover of the cyanogenic glycoside dhurrin has been observed in the intact *Sorghum* plant (*35*) cyanogenesis is normally prevented by spatial compartmentation of the cyanogenic substrate and the degrading enzymes (*3,26*). In the larvae of Z. *trifolii*, however, 1/3 of the total amount of linamarin and lotaustralin occurs in the hemolymph together with more than 80% of the linamarase activity (Table 1). Nevertheless, free hydrogen cyanide cannot be measured in fresh hemolymph from the larvae; cyanogenesis is not observed before several minutes have passed (Figure 2). Consequently, there is no degradation of the glucosides in the intact animal indicating that substrate and enzyme should be somehow compartmentalized under physiological conditions. Several attempts in order to destroy physical barriers such as vesicles by sonification and dissolution of the hemolymph in hypotonic medium did not increase cyanogenesis; finally, centrifugation of the hemocytes showed that no component of the cyanogenic system was present in the hemocytes, but both, substrate and enzyme, were present in the supernatant, the hemoplasma.

Figure 2 shows some experiments that provide some more insight into the system. For each experiment 10 µl hemolymph were diluted with 100 µl water, buffer or Ringer solution with and without substrate or enzyme and the rate of cyanide production was measured. The results show that pure, non-diluted hemolymph is slowly and weakly cyanogenic; when diluted with Ringer-solution or water cyanogenesis even decreases. When linamarin is added to the same amount of hemolymph diluted with Ringer-solution no increase of cyanogenesis can be observed. When, however, an equal amount of hemolymph diluted with Ringer-solution is supplied with a linamarase from *Hevea brasiliensis* (*33*) a quick and strong cyanogenesis results from the total liberation of glycosidically bound HCN. This indicates that the linamarase activity and not the availability of the substrate is the limiting factor of the system.

Chelating Agents Activate the Linamarase

Figure 2 also shows that an equal amount of hemolymph diluted with 150 mM McIlvaine buffer (Na-citrate/phosphate) shows a comparable kinetic of cyanogenesis indicating that this buffer system contains activating components. McIlvaine buffer is widely used for the isolation of enzymes; it was also used previously for the isolation of the linamarase of Z. *trifolii* (*30*), keeping the enzyme active and thus allowing it to be detected during the steps of purification.

When testing the components of McIlvaine buffer it became obvious that citrate rather than phosphate was the activating agent. In addition, ethylenediaminetetra acetatic acid sodium salt (EDTA) activated the linamarase in the same manner indicating chelating agents as being responsible for activation and, probably, metal ions for inhibition of the linamarase.

The elemental composition of the hemolymph was analyzed by inductively coupled plasma mass spectrometry (ICP-MS); Table 3 presents a part of the results showing those elements of more than 10^2 mg/L. In accordance with the literature (*36*) predominant cations are potassium, magnesium, calcium and sodium, two of which as ions form chelates, Mg^{++} and Ca^{++}, present in the ratio of ca. 2 : 1 in the hemolymph and at a total concentration of ca. 25 mM. When in a subsequent experiment the hemolymph was fully activated in 15 mM EDTA the activity of the linamarase decreased in dose dependent manner upon addition of either 1, 10 and 100 mM Mg^{++} or Ca^{++} ions (Figure 3 for Mg^{++}).

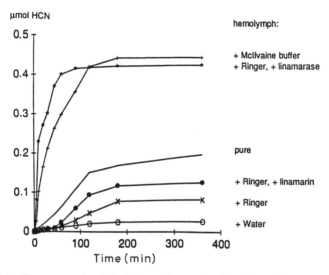

Figure 2. Cyanogenesis of the isolated hemoylmph under various conditions.

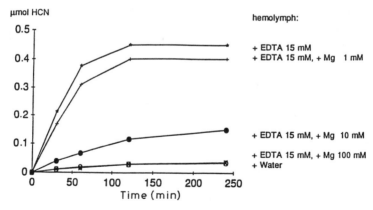

Figure 3. Inhibition of the linamarase in the hemolymph fully activated by EDTA at different concentrations of Mg^{++}.

Table III. Concentration of major elements in the hemolymph of larvae of *Zygaena trifolii* measured by ICP-MS

Element	mg/L	mM
K	0.83×10^3	21.3
Mg	0.44×10^3	18.3
Ca	0.30×10^3	7.5
Na	0.10×10^3	4.4
Cl	1.94×10^3	
S	1.24×10^3	
P	0.81×10^3	

Ca^{++} and Mg^{++} as Inhibitors of the Purified Linamarase

Up to this point all results were obtained using the crude hemolymph. The following experiments were undertaken with the isolated linamarase that was purified following mainly the protocol described in (*30*) but a last chromatographic step on concanavalin-A sepharose was added (Table 4) producing an enrichment of ca. 90-fold instead of ca. 50-fold obtained in (*30*). Binding on concanavalin-A sepharose indicates the enzyme is a glycoprotein. Because the isolation procedure used McIlvaine buffer the enzyme was free of Mg^{++} and Ca^{++} and, thus, in the active form. Experiments were carried out under non-chelating conditions either in water, adjusted to pH 6.2 to 6.5, or in an acetate buffer at pH 6.2 that is the physiological pH of the hemolymph.

Table IV. ß-Glucosidase from Larvae of *Zygaena trifolii*: Protocol of Purification

fraction	p-NPG nkat	p-NPG nkat/mg	factor	linamarin nkat	linamarin nkat/mg	factor
hemolymph	135	2.8	-	134	2.7	-
supernatant	133	3.1	1	133	3.1	1
G-25	137	3.2	1	136	3.2	1
DEAE	92	21	7.5	93	22	7.7
CM	87	198	71	81	185	69
Con-A	34.5	245	88	33	236	87

When the enzyme was thus incubated with concentrations of 1 to 150 mM of Mg^{++}/Ca^{++} in the physiological ratio of 2 : 1 a dose dependent inhibition was observed (Figure 4). At the physiological concentration of ca. 25 mM of both ions the linamarase had still 20% of its original activity in this experiment. When tested separately at 10 mM, Mg^{++} exhibited a stronger inhibitory activity than Ca^{++} (Figure 5); the slighly stronger inhibition of a 2 : 1 mixture at the same concentration may indicate a synergistic effect.

Further experiments (data not shown here) demonstrated a decrease of activity paralleling the decrease of the molarity of the acetate buffer; on the other hand the inhibitory effect of 10 mM Mg^{++}/Ca^{++} increases inversely to the buffer molarity. Thus at 150 mM buffer a 10 mM concentration of both ions showed a 50% inhibition, whereas at 25 mM acetate buffer the same concentration of both ions showed an almost 100% inhibition of the linamarase's activity. It was also shown that pre-incubation with Mg^{++}/Ca^{++} ions at 50 mM in acetate buffer at pH 6.2 for one hour at 20 °C leads to a stronger inhibition when compared to a non pre-incubated sample.

When the activity of a partially purified linamarase towards the endogeneous substrate linamarin was measured in the presence of two concentrations of the inhibitory ions it became evident that a lower concentration of Mg^{++} than of Ca^{++} was necessary to obtain a comparable decrease of activity (Figure 6 for Mg^{++}). These dara used for a Lineweaver-Burk plot indicate a noncompetitive or a mixed type of inhibition (*37*) for both ions (Figure 8 for Mg^{++}). The same data used in a Dixon plot to determine the inhibitor constant (*37*) resulted in values of 5 mM for Mg^{++} and 80 mM for Ca^{++} (Figure 7 for Mg^{++}); thus the inhibitor constant of Mg^{++} is clearly below its physiological concentration in the hemolymph of ca. 18 mM. Since the linamarase is a glycoprotein the ions may be bound to the carbohydrate portion probably via chelating groupings such as carboxyl and/or hydroxyl functions.

Conclusion

In conclusion these data strongly suggest at least a participation of Mg^{++} and Ca^{++} ions in the inhibition of the hemolymph's linamarase *in vivo*. This is, to our know-legde, the first clear cut observation of an inhibitory activity of these two alkaline earth metal ions on a ß-glucosidase involved in the metabolism of secondary con-stituents. Since many tests and protocols of purifications of ß-glucosidases are usually run with chelating buffers containing citrate such inhibition would not have been noticed for glucosidases in the past. We argue that, although the *in vitro* inhibition was not more than ca. 80%, the linamarase is totally inhibited *in vivo*. This makes sense because i. a continous turnover of linamarin and lotaustralin in a safe and not attacked larva would not be convenient from a teleological standpoint; ii. the detoxifying enzyme ß-cyanoalanine synthase that combines toxic cyanide with cysteine (or serine) to form ß-cyanoalanine is present at very low levels in the hemolymph (Table 1) where cyanide would be generated if the linamarase would be active.

It is, however, still unclear how the linamarase is activated after the larvae are injured and how pure hemolymph can, although weakly, be cyanogenic; it is also still not known why the larvae store 2/3 of their cyanoglucoside set in the cuticle and in a secretion stored in cuticular cavities (*38*) without linamarase activity (Table 1). The following may give some explanations:

The stability of chelates is strongly dependent on the pH; thus lowering the pH

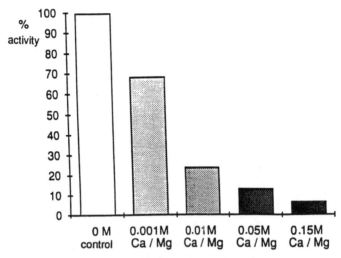

Figure 4. Inhibition of the purified linamarase by a 2 : 1 mixture of Mg^{++} and Ca^{++} at different concentrations.

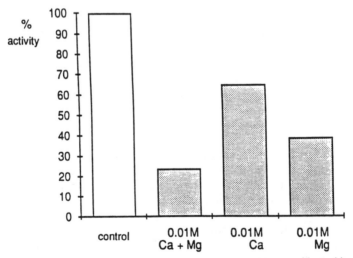

Figure 5. Inhibition of the purified linamarase by either Mg^{++}, Ca^{++} or a 2 : 1 mixture of both.

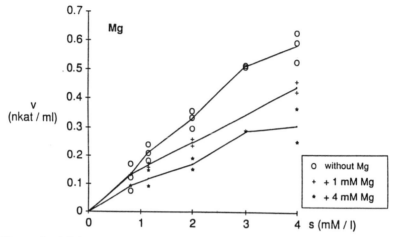

Figure 6. Michaelis-Menten plot of the linamarase in the presence of two concentrations of Mg^{++}.

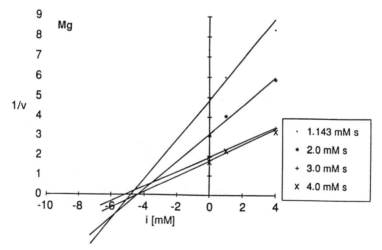

Figure 7. Dixon plot of the data from Figure 6.

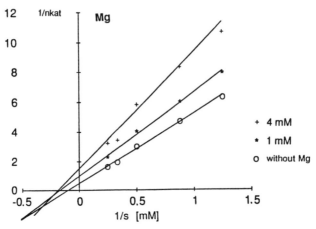

Figure 8. Lineweaver-Burk plot of the data from Figure 6.

will destroy chelates with Mg^{++} and Ca^{++} ions and activate the linamarase. In a preliminary experiment using Feigl-Anger test paper (39) to indicate cyanogenesis it could be shown that the hemolymph became more and more cyanogenic in non-chelating acetate buffer by lowering the pH from 6.2 (non cyanogenic) via 5.0 (non cyanogenic) and 4.3 (medium cyanogenic) to 3.6 (strongly cyanogenic comparable to McIlvaine at 6.2) (Figure 9).

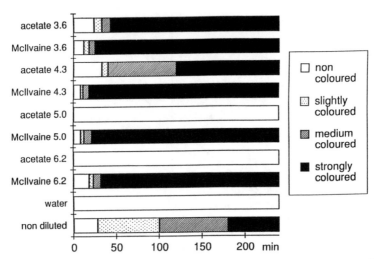

Figure 9. Cyanogenesis of the hemolymph in McIlvaine and acetate buffer at different pH for 240 min as tested with Feigl-Anger paper. The degree of blue colour indicates the degree of cyanogenesis.

This observation and the tissue distribution of linamarin and lotaustralin within the larvae (Table 1) led to our current model that the cyanogenic system of the larvae is mainly directed towards predators such as birds, lizards or small rodents which are able to damage heavily the larvae when eating them causing the hemolymph, secretion and cuticle to be mixed. Once a larva, thus injured, has reached the stomach of the insectivor the low pH there might activate the linamarase leading to effective cyanogenesis and, thus, to a repellent effect. However, we expect that this picture might by all means be corrected or changed by further work on this interesting system.

Acknowlegdement. We thank Dr. P. Quint (Münster) for the ICP-MS analysis, Dr. D. Selmar (Braunschweig) for providing a sample of linamarase from *Hevea brasiliensis*, Mr. Seipel (Büttelborn) for supplying the larvae, and Dr. A. Brinker for linguistic advice.

Literature cited

1. Schliemann, W. *Pharmazie* **1987**, *42*, 225-239.
2. Hösel, W., Conn, E. E. *Trends Biochem. Sci.* **1982**, *7*, 219-221.
3. Poulton, J. E. (1988) in *Cyanide Compounds in Biology*; Ciba Foundation; J. Wiley, Cichester, vol. 140. pp 67-80.
4. Schliemann, W. *Pharmazie* **1983**, *38*, 287-307.
5. Schliemann, W. *Biol. Rdsch.* **1983**, *21*, 333-347.
6. Baker, J. E. *Insect Biochem.* **1991**, *21*, 615-621.
7. Yu, S. J. *Insect Biochem.* **1989**, *19*, 103-108.
8. Hewitt, P. H., Retief, L. W., Nel, J. J. C. *Insect Biochem.* **1974**, *4*, 197-203.
9. Chararas, C., Chipoulet, J. M. *Comp. Biochem. Physiol.* **1982**, *72B*, 559-564.
10. Lindroth, R. L. *Insect Biochem.* **1988**, *18*, 789-792.
11. Reilly, C. C., Gentry, C. R., McVay, J. R. *J. Econ. Entomol.* **1987**, *80*, 338-343.
12. Chew, F. S. *ACS Symp. Ser.* **1988**, *380*, 155-181.
13. Nahrstedt, A. *Plant Syst. Evol.* **1985**, *150*, 35-47.
14. Sachdev-Gupta, K., Renwick, J. A. A., Radke, C. D. *J. Chem. Ecol.* **1990**, *16*, 1059-1067.
15. Brunet, P. C. J., Kent, P. W. *Nature* **1955**, *175*, 819-820.
16. Psarianos, C. G., Marmaras, V. J., Vournakis, J. N. *Insect* Biochem. **1985**, *15*, 129-135.
17. Chen, P. S., Neuweg, M. *Rev. Suisse Zool.* **1978**. *85*, 795-801.
18. Isobe, M., Kondo, N., Imai, K., Yamashita, O., Goto, T. *Agric. Biol. Chem.* **1981**, *45*, 687-692.
19. Rahbe, Y., Delobel, B., Bonnot, G. *Reprod. Nutr. Dev.* **1984**, *24*, 515-527.
20. Kramer, K. J., Hopkins, T. L., Ahmed, R. F., Mueller, D., Lookhart, G. *Arch. Biochem. Biophys.* **1980**, *205*, 146-155.
21. Ishizaki, Y., Umebachi, Y. *Sci. Rep. Kanazawa Univ.* **1980**, *25*, 43-52.
22. Ahmed, R. F., Hopkins, T. L., Kramer, K. J. *Insect Biochem.* **1983**, *13*, 641-645.
23. Davis, R. H., Nahrstedt, A. In *Comprehensive Insect Physiology, Biochemistry and Pharmacology*. Kerkut, G. A., Gilbert, L. I., eds. Pergamon Press, Oxford, 1985, Vol. 11; pp 635-54.

24. Nahrstedt, A. In *Cyanide Compounds in Biology*; CIBA Foundation; J. Wiley, Chichester, 1988, Vol. 140, pp 131-45.
25. Raubenheimer, D. *J. Chem. Ecol.* **1989**, *15*, 2177-2189.
26. Conn, E. E. *Planta med.* **1991**, *57*, S1-S9.
27. Nahrstedt, A., Davis, R. H. *Phytochemistry* **1986**, *25*, 2299-2302.
28. Selmar, D., Lieberei, R., Biehl, B., Conn, E. E. *Physiol. Plant.* **1989**, *75*, 97-101.
29. Mueller, E. *Untersuchungen zur Enzymologie der Cyanogenese und zum Metabolismus von Blausäure in den Larven von Zygaena trifolii (Esper, 1783) (Insecta, Lepidoptera)*, PhD-thesis, Univ. Muenster, 1992.
30. Franzl, S., Ackermann, I., Nahrstedt, A. Experientia **1989**, *45*, 712-718.
31. Pocsi, I., Kiss, L., Hughes, M. A., Nanasi, P. Arch. *Biochem. Biophys.* **1989**, *272*, 496-506.
32. Fan, T. W. M., Conn, E. E. *Arch. Biochem. Biophys.* **1985**, *243*, 361-373.
33. Selmar, D., Lieberei, R., Biehl, B. *Plant Physiol.* **1987**, *83*, 557-563.
34. Itho-Nashida, T., Hiraiwa, M., Uda, Y. *J. Biochem.* **1987**, *101*, 847-854.
35. Bough, W. A., Gander, J. E. *Phytochemistry* **1971**, *10*, 67-77.
36. Chapman, R. F. *The Insects 3. Edit.*; Havard Univ. Press, Cambridge Mass., 1982, pp 806-809.
37. Dixon, M., Webb, E. C. *Enzymes 3. Edit.*, Academic Press New York, 1979, pp 332 ff.
38. Franzl, S., Naumann, C. M., Nahrstedt, A. *Zoomorphology* **1988**, *108*, 183-190.
39. Brinker, A. M., Seigler, D. S. *Phytochem. Bull.* **1989**, *21*, 24-31.

RECEIVED March 2, 1993

Chapter 10

Function and Variation of the β-Glucosidase Linamarase in Natural Populations of *Trifolium repens*

P. Kakes

Free University, de Boelelaan 1087, 1081 HV Amsterdam, Netherlands

Cyanogenesis is the production of HCN by plants, and in some cases, by animals. Cyanogenesis is only apparent after damage of the cyanogenic organs or tissues of the plant. The mechanism responsable for cyanogenesis is hydrolysis of one or more cyanogenic substrates by a ß-glucosidase. The function of cyanogenesis is best studied in species that are polymorphic for cyanogenesis. *Trifolium repens* (white clover) is polymorphic for both the cyanogenic substrates and the ß-glucosidase linamarase. This situation makes it possible to study the function of linamarase in plants with and without the natural substrates. The most obvious function of cyanogenesis is defense against herbivores. However I was able to show that linamarase has no function in the deterrence of slugs and snails by the cyanogenic glucosides present in white clover. It is possible that linamarase has a function in the deterrence of other herbivores, but the evidence for this is weak. There is much variation in linamarase content in natural populations of white clover. The significance of this variation is discussed within the framework of the cost-benefit hypothesis.

In white clover (*Trifolium repens*) cyanogenesis, i.e. the production of HCN is caused by the action of the cyanogenic β-glucosidase linamarase on two cyanogenic substrates: linamarin and lotaustralin (*1*) (See Figure 1)

The substrates occur in varying proportions in leaf cells, presumably in the vacuole. Linamarase is an apoplastic enzyme, occurring mainly in the walls of the epidermal cells of the leaves (*2*). Linamarase activity in other organs of *T.repens* is low or undetectable (*3*), (Kakes, unpublished). The compartmentalisation of enzyme and substrate (Figure 2) insure that no detectable amount of HCN is produced in intact plants.

As in other plants, the cyanogenic system in white clover is generally looked upon as one of the mechanisms of defense against herbivores. There is ample evidence that cyanogenesis protects plants against generalized herbivores (but not specialized ones) but of course this does not exclude other functions of the cyanogenic system or of its components.

0097–6156/93/0533–0145$06.00/0

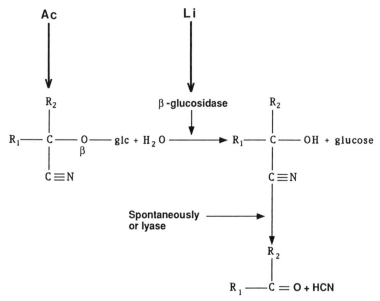

Figure1. General scheme of the hydrolysis of cyanogenic substrates. The effects of the genes Ac and Li in *T. repens* are indicated

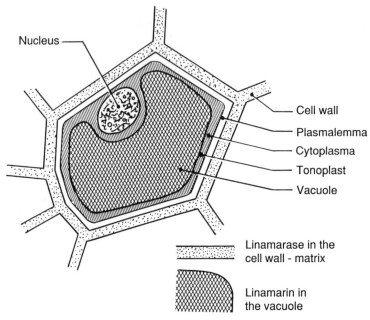

Figure 2. The localisation of linamarase and the presumed localisation of the cyanogenic substrates.
(Reproduced with permission from *Vakblad voor Biologen* **1986**, *66*, 13. Copyright 1986.)

T. repens is is a species that is well suited for an investigation of the cyanogenic system as most populations contain cyanogenic and acyanogenic plants. In population studies only long established fields or natural stands should be used, to prevent contamination with cultivars. Fortunately gene flow is very limited in *T.repens* (4).The polymorphism is caused by variation in two genes: *Li* regulates the presence/absence of linamarase and *Ac* the presence/absence of the two cyanogenic substrates. The variation in *Ac* and *Li* produce four so called cyanotypes that can be simply distinguished (Figure 3). Both genes have a dose effect: in plants with two active alleles twice as much product is produced, compared to plants with one active allele (5, 6). Genetical tests show that *Ac* and *Li* are unlinked. In this paper I will discuss the variation in linamarase and the possible functions of this enzyme in natural populations of white clover and some related species.

Properties of Linamarase

Linamarase in *T.repens* is homodimer with a subunit molecular mass of 62.000 Dalton. It is a glycoprotein and can represent up to 5% of the soluble proteins in young leaves (7). As mentioned before it is localized in the cell wall. In plants homozygous for the *li* allele no linamarase activity is found and also no anti-linamarase cross reactive material can be detected (8, 9). In genotypes and organs without linamarase activity a low but distinct β-glucosidase activity can be found with artificial substrates. This activity may be due to a β-glucosidase (or β-glucosidases) not under the control of the Li-gene (10). Kinetic studies show that the white clover linamarase has a broad substrate specificity, depending on the aglycone and on the type of glucosidic linkage (11). The two known substrates in *T.repens* are both split by linamarase.

Distribution of the Li-alleles in European Populations

There are at least three alleles of Li differing in linamarase activity (6) but only plants homozygous recessive for the null allele (*lili*) can be simply distinguished from the other genotypes. It has long been known that in Europe the frequency of the null allele is low in populations around the Mediterranean Sea and gradually increases to the North (12). In Scandinavia and northern Russia, where the species has its northern limit, very few plants with linamarase have been found. A similar pattern is found when we study mountain populations: the higher the altitude, the lower the frequency of linamarase containing plants. The decrease is however not gradual: up to 600m high frequencies of the dominant allele prevail. Over 600m the range of frequencies broadens, with a definite dowward trend. As this pattern is observed both in the Cevennes in South-eastern France (13) (Figure 4) and in the Western Pyrenées (Kakes, unpublished) it is not likely to be a chance event.

   The corresponding latitudinal and altitudinal clines suggest that low temperatures, especially low winter temperatures, in some way determine the large scale distribution of *Li*. It should be noted that the frequency of Ac follows a similar pattern. Although in local populations almost every frequency of the four cyanotypes may be observed the large scale distributions of *Li* and *Ac* are correlated. Local populations often show linkage disequilibrium, i.e.the phenotype having both enzyme and substrate is found more often than expected on the basis of independent assortment of the alleles. (14), (15)

Studies of the Function of the Cyanogenic System, in Particular of Linamarase

The occurrence in one population of plants with and without linamarin or linamarase poses a problem: If we assume that cyanogenesis has a function, why is it that most

Added:	Water	Substrate	Enzyme
Genotype:			
Ac-Li-	+	+	+
Ac-lili	-	-	+
acacLi-	-	+	-
acaclili	-	-	-

Figure 3. The four cyanotypes that can be distinguished with semiquantitative tests.

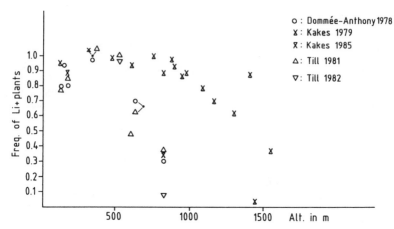

Figure 4. The distribution of the Li alleles in South-Eastern France. Results of different investigators and years have been combined. All data are from Till et al (13)

(Reproduced with permission from reference 13. Copyright 1988 Gauthier–Villars.)

populations contain individuals that do not have this system, or have an incomplete one?

The variation in chemical defense in general has given rise to a considerable body of hypotheses and I will discuss only briefly one of them: The cost-benefit hypothesis (see also *16*). It is postulated that the production and maintenance of a defensive system has a cost for the plant. Natural selection will favour such a system only so long as the benefits are greater than the costs.

To study cost and benefit of cyanogenesis I produced a repeated backcross generation segregating for the genes *Ac* and *Li*. The different cyanotypes were followed from the germination of the seeds until seedset of the adult plants in the second year. The results (*16*) showed that for *Ac* the cost-benefit hypothesis works well: Plants that possess linamarin/lotaustralin (*Acac*) are less attacked by slugs and snails than acac plants, both in the seedling stage and as adult plants. Table I shows the result of the seedling experiment: the difference with the non grazed control (not shown) is significant for *Ac* but not for *Li*. Figure 5 demonstrates the effect of grazing on adult plants.

Clearly the protection of seedlings by the presence of cyanogenic glucosides has a selective advantage, as we know that under natural conditions slugs and snails destroy a large proportion of the seedlings. The cost of defense appears when the plants reproduce, as shown in figure 6. *Acac* plants produce only half of the flowers and seeds compared to plants without linamarin/lotaustralin (*acac*).

For linamarase the situation is different: Our results clearly show that linamarase is not necessary for the deterrence of slugs and snails. This result, although not obtained earlier, is not surprising, as the enzyme mixture in the gut of snails has a high linamarase activity (*17*), (Kakes, unpublished). It is possible that linamarase is necessary for the defense against other herbivores lacking linamarase activity in their guts, but the evidence for selective eating by organisms other than molluscs is weak. In a recent review Hughes (*3*) lists 10 papers in which selective eating of acyanogenic *T.repens* is reported. These 10 papers give 14 species or species groups. Ten of these are molluscs (slugs and snails), the remaining 4 are insects. However the two insect reports are far from convincing. In the first report (*18*) different reproduction of the aphids on cyanogenic and acyanogenic plants could well explain the data given by the authors. In the case of the other two insect species (*19*) not cyanotypes but commercial varieties differing in frequency of cyanogenic plants were compared. In none of the studies except that of Kakes (*16*) were the effects of *Ac* and *Li* studied separately.

Several other differences between plants with and without linamarase have been reported in the literature: linamarase containing plants have bigger leaves and better survival as young plants (*20*), shorter stolons (*21*) and higher rootgrowth (*22*) compared to plants without the enzyme. These characteristics cannot be easily related to the known properties of the enzyme and possibly constitute the effects of other genes linked to *Li*. The frost resistance of *lili* plants, once thought of as the reason of the geographical clines, has not been confirmed by Kakes: of 80 plants comprising the four cyanotypes in equal frequencies four individuals died in the comparatively severe winter of 1985/1986. Two of the survivors were *AcacLili,* and the other two were *acaclili*. (See *16* for full experimental details)

So we are here in a rather paradoxal situation: Although biochemical evidence suggests that linamarin and linamarase form a coördinated system, we lack direct experimental evidence that they serve a coördinated ecological function. There are several ways out of this dilemma: the first is that we may be looking in the wrong direction: linamarase could have a function that is related to the metabolism of the cyanogenic glucosides, but not to chemical defense. The second is that linamarase has indeed a function in the defensive system, but that we have as yet failed to study the right herbivores under the right conditions. The third possibility is that although linamarase has served some function in the past, it does no longer do so, at least not in the majority of extant white clover populations. This may not seem obvious, but we

150   β-GLUCOSIDASES: BIOCHEMISTRY AND MOLECULAR BIOLOGY

Table I. Seedling Experiment Results

| | Cyanotypes: | | |
	Lili	*lili*	Total
Acac	30	20	50
acac	5	5	10
Total	35	25	60

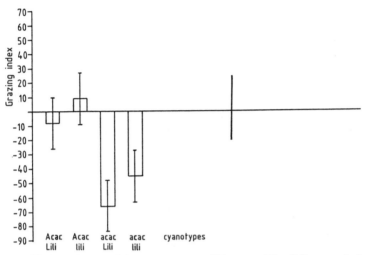

Figure 5. Grazing index of the cyanotypes of *T.repens.* The difference in leaf area before and after a week of grazing by the snail *Helix aspersa* is expressed on the Y-axis as percentage of the difference in leaf area of a non grazed control. As the latter difference was positive, a negative value of the grazing index means that more leaf   was removed by grazing than added by growth. The effect of *Ac* is significant but that of *Li* is not.

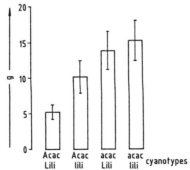

Figure 6. Mean dry weight of the ripe flower heads of the cyanotypes of *T. repens*. The effect of *Ac* is significant but that of *Li* i s not.
Cyanotypes of the survivors of 178 seedlings of *T.repens* grazed for 18 days by *Helix aspersa* and *Cepaea nemoralis*. The proportion of *Acac - acac* plants differs significantly from a non-grazed control, but that of *Lili -lili* not. See also (*16*).

must realize that of the two species ancestral to *T.repens* one, *T.nigrescens*, with a southerly distribution is monomorphic for cyanogenesis, i.e. all individuals contain linamarase, whereas in the other species, *T.occidentale*, no plants with linamarase have been found to date, although the majority of plants contain linamarin/lotaustralin (*23*, Kakes, unpublished).

So far I have discussed genetic variation in linamarase content. However there is also non-genetic variation. The expression of the *Li*-gene is dependent on the environment (*24, 25*). Especially at high temperatures the expression of certain genotypes is strongly depressed so that they are difficult to distinguish from *lili* plants. We have developed a method for the quantitative determination of linamarase in single leaflets of white clover. We are planning to use this method to study the effects of environmental factors on the expression of *Li* under natural conditions. A pilot experiment, to study the validity of the method, has unexpectedly shown that the leaflets of one leaf may show a two to threefold difference in linamarase content. It is difficult to believe that these differences are caused by the environment. It rather looks like an endogenous mechanism, although we will have to do more experiments to reach a firm conclusion. I mention these results for the following reason: in order to understand the function and variation of linamarase we must look upon the system from the point of view of the plant on one hand and from the point of view of the herbivores on the other. From the point of view of the plant it is irrelevant if some part of it is eaten by herbivores so long as survival and reproduction are not impaired.

The herbivores, especially the smaller ones, see the world as a mosaic of more and less palatable plant parts. White clover adds to the complexity of their world both by genetic means (cyanogenic and acyanogenic clones growing intermingled) as by epigenetic means (environmental and endogenous variation). It is this complexity that prevents a herbivore from adapting to one specific type of plant. In a recent paper Till-Bottraud and Gouyon (*26*) show that under certain conditions natural selection will favour a mixture of cyanogenic and acyanogenic plants or leaves. It may well be that variation is an essential part of the natural function of linamarase.

Conclusions

We have seen that from a biochemical point of view linamarase has a well defined function: it is the only plant enzyme that effectively hydrolyzes the cyanogenic substrates. No other natural substrates have been discovered as yet, at least not in white clover. The evidence for its ecological function is less clear. Certainly it is not an essential enzyme: as we saw before it is absent from many plants and some populations and there is even one species closely related to *T.repens* in wich the substrates are present and the enzyme is presumably lacking. On the other hand there is circumstantial evidence of its function: That evidence is of a statistical nature: the correlated distributions of *Ac* and *Li*, both on a local and on a geographical scale.

There is no direct experimental evidence that linamarase is necessary for the protection against herbivores. However, we have isolated observations of selective eating of *lili* plants by waterfowl. Although such observations do not carry much weight in themselves, they strenghten the conclusion from statistical data that there is indeed a function for linamarase. In any case, the coöcurrence of plants with and without linamarase in most populations suggests that variation in linamarase content may be the evolutionary optimum for white clover in most populations.

Literature cited

1.   Melville, J. ; Doak, B.W. *N.Z.J. Sci. Techn.* **1940**, *22B*, 67-71.
2.   Kakes, P. *Planta* **1985**, *166*, 156-160.
3.   Hughes, M.A. *Heredity* **1991**, *66*, 105-115.

4. Gliddon, C., Saleem, M. In: Genetic Differentiation and Dispersal in Plants P. Jacquard et al Eds.; Springer: Berlin, **1985**, 293-308.
5. Hughes, M.A. ; Stirling, J.D. *Euphytica* **1982**, *3*, 477-483.
6. Maher, E.P.; Hughes, M.A. *Biochem. Gen.* **1973**, *8*, 13-15.
7. Hughes, M.A.; Dunn, M.A. *Plant Mol. Biol.* **1982**, *1*, 169-181.
8. Hughes, M.A.; Dunn, M.A; Pearson, J.R. *Heredity* **1985**, *55*, 387-391.
9. Kakes, P.; Eeltink, H. *Z. Naturforsch.* **1985**, *40c*, 509-513.
10. Boersma, P.; Kakes, P.; Schram,.A.W. *Acta Bot. Neerl.* **1983.***32*, 39-47.
11. Pócsi, I., Kiss, L.; Hughes, M.A.; Nánási, P. *Arch. Biochem. Biophys.* **1989**, *272*, 496-506.
12. Daday, J. *Heredity* **1954**, *8*, 61-78.
13. Till, I.; Kakes, P.; Dommée, B. *Oecol. Plant* **1988**, *9*, 393-404.
14. Ennos, R.A. *Genet. Res.* **1982**, *40*, 65-73.
15. Kakes, P. *Acta Bot. Neerl.* **1987**, *36*, 59-69.
16. Kakes, P. *Theor. Appl. Genet.* **1989**, *77*, 111-118.
17. Dirzo, R. ; Harper, J.L. *J. Ecol.* **1982**, *70*, 101-117.
18. Dritschilo, W; Krummel, J.; Pimentel, D. *Heredity* **1979**, *42*, 49-56.
19. Mowat, D.J.; Shakeel, M.A. *Grass Forage Sci.* **1988**, *43*, 371-375.
20. Ennos, R.A. *Heredity* **1981**, *46*, 127-132.
21. Kakes, P. **1990**. *Euphytica 48*, 25-43.
22. Dommée, B.; Brakefield, P.M.; Macnair, M.R. *Oecol. Plant* **1980**, *1*, 367-370.
23. Gibson, P.B.; Barnett, O.W; Gillingham, Y.T. *Crop Sci.* **1972**, *12*, 708-709.
24. Waal, R. de. Thesis Landbouwhogeschool, Wageningen, The Netherlands **1942**.
25. Till, I. *Heredity* **1987**, *59*, 265-271.
26. Till-Bottraud, I.; Gouyon, P.H. *Am. Nat.* **1992**, *139*, 509-520.

RECEIVED January 12, 1993

Chapter 11

Molecular Genetics of Plant Cyanogenic β-Glucosidases

M. A. Hughes

Department of Biochemistry and Genetics, University of Newcastle upon Tyne NE2 4HH, United Kingdom

Three higher plant β-glucosidase genes have been cloned as cDNA. Analysis of the deduced amino acid sequence of the cyanogenic β-glucosidase from white clover and from cassava and a non-cyanogenic β-glucosidase from white clover reveals considerable homology between the three genes and also with several cloned prokaryotic β-glucosidase enzymes. Both of the plant cyanogenic β-glucosidases are post translationally modified by signal peptide cleavage and glycosylation. The white clover cyanogenic enzyme has an extracellular location but the cassava enzyme is intracellular, being synthesised and subsequently located specifically within the latex vessels of this plant.

The production of hydrocyanic acid (HCN) by higher plants depends upon the co-occurrence of a cyanogenic glucoside and catabolic enzymes. In white clover (*Trifolium repens* L), two related cyanoglucosides are produced, 1-cyano-1-methylethyl ß-D-glucopyranoside (linamarin) and R-1-cyano-1-methylpropyl ß-D-glucopyranoside (lotaustralin) *(1)*. These cyanoglucosides are also found in *Manihot esculenta* Cranz (cassava) *(2)*, *Linum usitatissimum* L. (flax) *(3)*; *Phaseolus lunatus* L. (lima bean) *(4)*, *Hevea braziliensis* L. (rubber) *(5)* and *Lotus* species *(6)*. In general cyanogenic glucosides are broken down sequentially following tissue damage, which results in disruption of the compartmentation of the stored glucosides and a ß-glucosidase (linamarase). Hydrolysis of the cyanoglucosides by a ß-glucosidase produces glucose and a hydroxynitrile, which in many plants is subsequently broken down by a hydroxynitrilase to a ketone and HCN (Figure 1). Since all of the above species produce the same two cyanoglucosides, it follows that the cyanogenic ß-glucosidase produced by these plants has specificity for the same aglycone structure.

Cyanogenic Polymorphism in White Clover

The first report of the cyanogenic polymorphism in white clover was by Armstrong *et al (7)*. Since this report, the polymorphism, that is, the existence of both cyanogenic and acyanogenic plants in natural populations, has been the subject of a large number of ecological studies. Modified dihybrid Mendelian segregation ratios, in progeny scored for the production of HCN, were demonstrated in the 1940s by a group working in New Zealand *(8)*. Acyanogenic plants fall into three phenotypic classes; plants which produce no cyanoglucosides, plants which produce no cyanogenic ß-glucosidase and plants which produce neither. The presence of cyanoglucoside is determined by alleles of a single locus (*Ac*), whilst the presence of cyanogenic ß-glucosidase activity is determined by alleles of an unlinked locus (*Li*).

Table I   The Effect of Genotype on Cyanogenic ß-glucosidase Activity in *Trifolium Repens* L. (White Clover)

Genotype	Mean Specific Activity*
Li Li	0.258
Li Li	0.235
Li li	0.093
Li li	0.087
Li li	0.120
li li	0.000
li li	0.000

*μmoles linamarin hydrolysed/mg protein/min

A number of biochemical studies have been carried out to characterise non-functional alleles at the *Li* locus. Incomplete dominance has been shown at the biochemical level *(9)*. Table I illustrates this by showing the activity of crude leaf extracts from white clover plants of different genotype measured against the natural cyanogenic substrate, linamarin. Heterozygous *Lili* plants thus have intermediate levels of cyanogenic ß-glucosidase activity. Antibodies raised to purified cyanogenic ß-glucosidase have been used to quantify amounts of cyanogenic ß-glucosidase protein produced in white clovers plants of different genotype *(10)*. Figure 2 shows that affinity purified cyanogenic ß-glucosidase antibodies reveal no detectable antigen in plants homozygous for the non-functional *Li* alleles (*lili*). The *lili* plants with no cyanogenic ß-glucosidase activity therefore contain no cyanogenic ß-glucosidase protein. A study of variant forms of white clover, which produce reduced levels of cyanogenic ß-glucosidase activity, has shown that these reduced levels of activity are due to

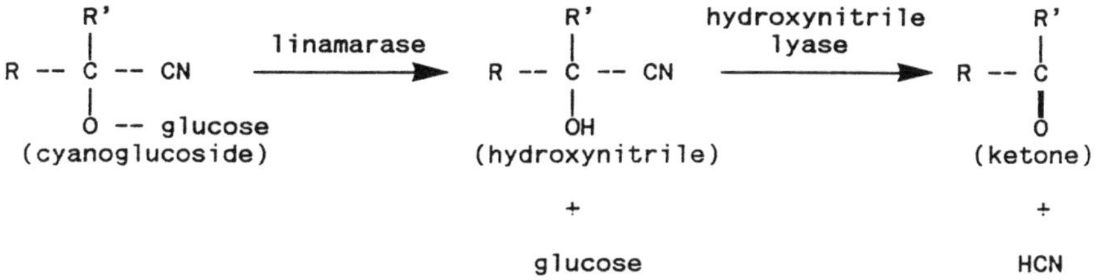

$$R -- \overset{\overset{\displaystyle R'}{|}}{\underset{\underset{\displaystyle (cyanoglucoside)}{O -- glucose}}{C}} -- CN \xrightarrow{\quad \text{linamarase} \quad} R -- \overset{\overset{\displaystyle R'}{|}}{\underset{\underset{\displaystyle (hydroxynitrile)}{OH}}{C}} -- CN \xrightarrow[\quad]{\quad \text{hydroxynitrile} \atop \text{lyase} \quad} R -- \overset{\overset{\displaystyle R'}{|}}{\underset{\underset{\displaystyle (ketone)}{O}}{C}}$$

$$+ \qquad\qquad\qquad +$$

$$\text{glucose} \qquad\qquad \text{HCN}$$

Figure 1. The release of HCN from the cyanoglucosides; R = CH₃, R'
= CH₃ linamarin; R = CH₃ , R' = C₂H₅ lotaustralin; linamarase =
cyanogenic ß-glucosidase.

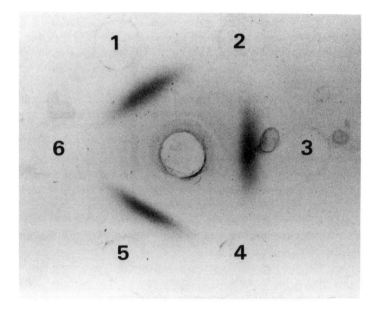

Figure 2. Double diffusion test using affinity purified antibodies raised to
active cyanogenic ß-glucosidase; Wells 1,3,5, leaf extract genotype *Li Li*;
Wells 2,4,6, leaf extract genotype *li li*; Centre well, cyanogenic ß-
glucosidase antibodies; Immunoprecipitate stained with Coomassie blue.

reduced rates of synthesis of the enzyme in developing leaves *(9,10)*. This character is determined by a genetic element which lies within 4 map units of the *Li* locus. Thus all the available evidence indicates that non-functional alleles at the *Li* locus result in reduced or zero cyanogenic ß-glucosidase synthesis and have characteristics of mutations in a cis-acting regulatory region.

Cloning the Cyanogenic ß-glucosidase from White Clover and Cassava

The cyanogenic ß-glucosidase from white clover is a homodimer with a subunit molecular weight of Mr 62,000 *(11)*. Concanavalin A affinity chromatography and analysis of endoglycosidase H (endo-ß-N-acetylglucosaminidase H) digestion of the enzyme using SDS-PAGE (Figure 3) shows that this ß-glucosidase is glycosylated, having high-mannose-type N-asparagine-linked oligosaccharides *(11,12)*. Glycoproteins of this nature are glycosylated in the lumen of the endoplasmic reticulum (ER) and translocation into the ER is usually accompanied by cleavage of a signal peptide from the N-terminus of the nascent polypeptide. Since these processes are co-translational, the only detectable product of *in vivo* synthesis is the fully processed glycoprotein. Figure 4 shows the results of a study of *in vivo* cyanogenic ß-glucosidase synthesis in white clover *(11)*. The antibiotic tunicamycin, which prevents N-acetyl-asparagine glycosylation, has been widely used to detect precursor forms of glycoproteins. However, Figure 4 demonstrates that, although tunicamycin reduces the synthesis of the ß-glucosidase, there is no evidence for the accumulation of precursor molecules in white clover leaf tissue, suggesting an intimate relationship between post-translational processing and translation of the mRNA.

Information obtained from *in vivo* radiolabelling experiments indicated that maximum levels of cyanogenic ß-glucosidase synthesis occur in young leaves of white clover plants (genotype, *LiLi*). mRNA was extracted from this material (*LiLi*) and from equivalent leaves of *lili* plants, which produce no cyanogenic ß-glucosidase protein. Figure 5 shows the results of immunoprecipitation of the *in vitro* translation products from these mRNA extractions *(11)*. A Mr 59,000 product was immunoprecipitated from *LiLi* mRNA reactions but not from *lili* mRNA reactions. Furthermore, addition of unwashed dog pancreas microsomes (which will process the nascent polypeptide) to the *in vitro* translation reaction, resulted in the production of a Mr 62,000 protein. This indicates that the nascent polypeptide is Mr 59,000 and by analogy with other systems, proteolytic removal of a N-terminal peptide and subsequent glycosylation in the ER will combine to produce a Mr 62,000 protein. The size of the unglycosylated proteolytic product is not known.

The information of cyanogenic ß-glucosidase synthesis in *LiLi* and *lili* plants was used to devise a strategy to clone the white clover cyanogenic ß-glucosidase. A cDNA library was made from *LiLi* young leaf mRNA in the vector pBR322. The assumption was made that the lack of active cyanogenic ß-glucosidase mRNA in *lili* plants was due to the absence of mRNA molecules. A differential screen using *LiLi* and *lili* cDNA probes identified 70 colonies which gave a stronger signal with the *LiLi* cDNA probe. Sixteen of these clones were

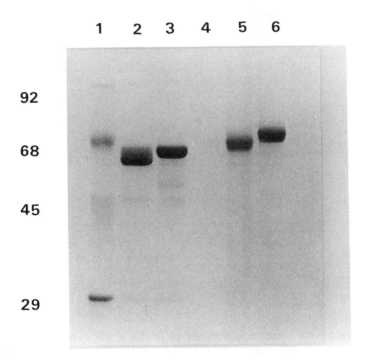

Figure 3. SDS-PAGE of purified cyanogenic ß-glucosidase from cassava and white clover: Effect of endoglycosidase H on relative molecular mass; Well 1, molecular weight markers, Mr x 10^{-3}; Well 2, endoglycosidase H digestion of white clover enzyme; Well 3, untreated white clover enzyme; Well 4, endoglycosidase H; Well 5, endoglycosidase H digestion of cassava enzyme; Well 6, untreated cassava enzyme.

(Reproduced with permission from ref. 12.)

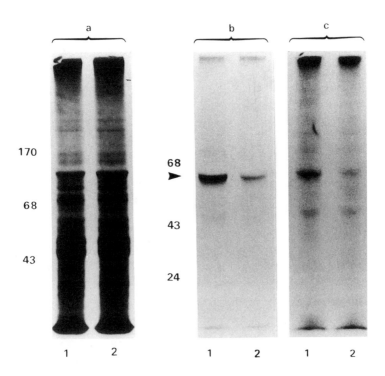

Figure 4. The effect of tunicamycin on *in vivo* synthesis of cyanogenic ß-glucosidase in 3-day-old white clover seedling cotyledons (variety Olwen). Fluorographs of total soluble proteins (a) and ß-glucosidase immunoprecipitated using antibodies raised to denatured cyanogenic ß-glucosidase (b and c) and analyzed by SDS-PAGE. (a) Crude extract supernatants showing total soluble proteins synthesized in the absence (lane 1) and presence (lane 2) of tunicamycin. (b) Immunoprecipitates from crude extract supernatants boiled with 0.1% SDS; lane 1, no tunicamycin, lane 2, 100 μg ml^{-1} tunicamycin. (c) Immunoprecipitates from resuspended cell debris boiled with 0.1% SDS; lane 1, no tunicamycin; lane 2, 100 μg ml^{-1} tunicamycin.   The positions of molecular weight markers (M_r X 10^{-3}) and cyanogenic ß-glucosidase protein (▶), run on the gel and stained with Coomassie blue, are indicated. (Reproduced with permission from ref. 11)

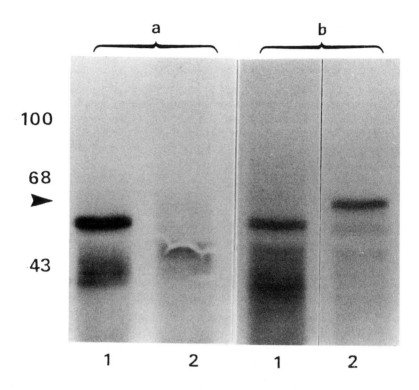

Figure 5. *In vitro* synthesis and processing of white clover cyanogenic ß-glucosidase in the wheat germ cell-free system (radiolabeled with [^{3}H]leucine). (a) *In vitro* translation products of mRNA from a cyanogenic ß-glucosidase-positive plant Eg, genotype *Li Li* (lane 1) and a cyanogenic ß-glucosidase-negative plant RD1, genotype *li li* (lane 2), immunoprecipitated with antibodies to denatured cyanogenic ß-glucosidase. (b) *In vitro* translation products of mRNA from plant Eg without dog pancreas microsomes (control, lane 1) and with the addition of unwashed dog pancreas microsomes (lane 2). Immunoprecipitates analyzed by SDS-PAGE and fluorography. The positions of molecular weight markers (M_r X 10^{-3}) and cyanogenic β-glucosidase protein (▶), run on the gel and stained with Coomassie blue, are indicated. (Reproduced with permission from ref. 11.)

screened by hybrid-select translation and one clone (TRE36) was judged to contain the cyanogenic ß-glucosidase by the following criteria : (1) the clone hybridised with an mRNA from *LiLi* leaves which gave a Mr 59,000 polypeptide as the major *in vitro* translation product; (2) this polypeptide was immuno-absorbed by affinity purified antibodies; (3) when the clone was used to hybrid-select mRNA from *lili* leaves, no Mr 59,000 polypeptide product was detected (Figure 6) *(13)*. Comparison of the insert size and the size of the cyanogenic ß-glucosidase transcript indicated that the clone TRE36 did not contain a full length cDNA.

A new white clover cDNA library was constructed in the lambda vector, gt10, and TRE36 was used as a probe to isolate further clones *(14)*. Clones with inserts bigger than 1000 bases were subcloned into pBluescript KS and the sequencing vector M13mp18. Restriction site analysis indicated that two classes of cDNA clone had been isolated by this procedure. Type I, which was identical to the original clone TRE36 and Type II (e.g. TRE104) which differed in having only a single Hind III site and no Kpn I or Ssp I sites.

The white clover ß-glucosidase clones were used as heterologous probes to identify cassava cyanogenic ß-glucosidase clones. Measurements of cyanogenic ß-glucosidase activity during the first 12 days of cassava seed germination showed that there was a rapid increase in total activity per seedling during germination. Most of the enzyme was found in the cotyledons and hypocotyl. A cassava cDNA library was made in the lambda vector gt10, using EcoR1/Not 1 adaptors, from mRNA extracted from 10-day old cotyledons *(12)*. The library was screened with a 704-base Ssp I fragment of the white clover ß-glucosidase cDNA clone, TRE361, which had been shown to include a region conserved in a number of ß-glucosidase genes *(14)*. Six clones were recovered and subcloned into M13mp18 and pBluescript KS. Subsequent sequence analysis showed that all six clones represented the same mRNA. Northern blot analysis, using one of the largest clones, CAS5 (1705 bases), revealed a transcript of 1,800 bases and the clone was considered virtually full length.

Analysis of Cyanogenic ß-glucosidase Clones

The nucleotide and derived amino acid sequence of the longest Type II clone (TRE104, 1,690 bases) is shown in Figure 7. This sequence contains three base changes from the previously published sequence *(14)* which extends the open reading frame at the 3' end of the clone. The N-terminal amino acid sequence and a 23 residue internal peptide sequence of the white clover cyanogenic ß-glucosidase were determined and found to be present in the deduced amino acid sequence of TRE104. The N-terminal sequence is missing from the deduced amino acid sequence of the Type I clone, TRE361 and the internal peptide has 4 amino acids which are different, although 3 of these are equivalent substitutions. There is considerable homology between TRE104 and TRE361 (67% deduced amino acid sequence homology) although the homology is not uniformly distributed along the sequences. The putative N-glycosylation sites and polyadenylation signal have been marked on the TRE104 sequence (Figure 7).

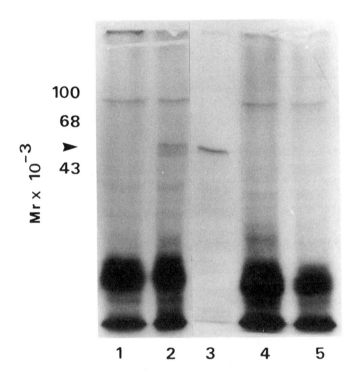

Figure 6. SDS-PAGE of *in vitro* translation products obtained with hybrid selected RNA from white clover developing leaf tissue. Well 1, *li li* tissue, total translate of RNA hybrid selected with pTRE36 plasmid DNA; Well 2, *Li Li* tissue, total translate of RNA hybrid selected with pTRE36 plasmid DNA; Well 3, anti-denatured-cyanogenic ß-glucosidase immunoprecipitate from total translate in Well 2; Wells 4 and 5, *Li Li* tissue, total translate of RNA hybridised with plasmid DNA from two control clones. The positions of Coomassie blue stained molecular weight markers and the 59,000 Mr cyanogenic ß-glucosidase nascent polypeptide (▶) are indicated. (Reproduced with permission from ref. 13).

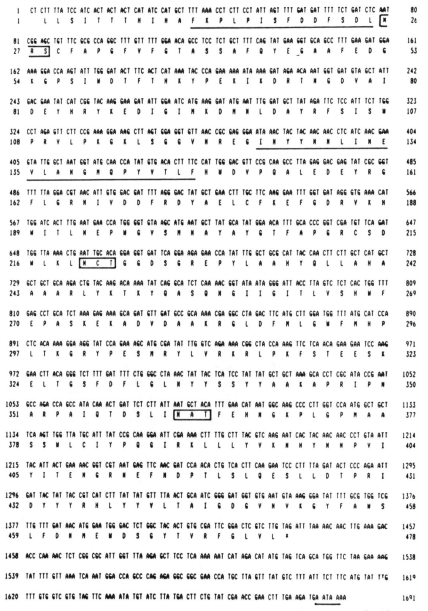

Figure 7. Nucleotide sequence and deduced amino acid sequence of white clover cDNA clone TRE104 (Type II). Amino acids homologous to known peptide sequence are underlined; putative N-glycosylation sites are boxed; putative polyadenylation signals on the DNA are underlined. (Adapted from ref. 14).

Table II   Northern Blot Analysis of White Clover Leaf RNA
from Different Genotypes

	Probe*	Genotype Li Li	li li
Gel I	TRE36	6,636	1,461
	TRE104	15,626	0
Gel II	TRE104	26,322	0
	TRE36	6,344	1,104

Peak integral values determined on a Chromoscan 3 densitometer.
10μg total RNA loaded per well.   * in order used.
TRE36 is 1.73K6 fragment of the non-cyanogenic β-glucosidase
cDNA, TRE361 *(13)*; TRE104 is complete cDNA of
the cyanogenic β-glucosidase *(14)*.

In order to confirm TRE104 as the cDNA for the cyanogenic ß-glucosidase Northern blots of total RNA extracted from young leaves of *LiLi* and *lili* plants were probed with clone TRE36 (Type I) and clone TRE104 (Type II) and washed at high stringency *(14)*.   Table 2 shows peak integral values of a densitometric analysis of this Northern blot analysis.   Both clones show a larger signal in the cyanogenic ß-glucosidase-producing genotype (*LiLi*) compared with the non-ß-glucosidase-producing genotype (*lili*).   However, clone TRE104 reveals no transcript in the *lili* plant extract whilst TRE36 gave a signal in both filters. The result suggests that TRE36 is recognising a ß-glucosidase mRNA present in *lili* plants which is not detectable by TRE104.   Both probes hybridised with mRNA of the same size, estimated as 2,000-1,900 bases *(14)*.   These results are consistent with Type I clones (eg TRE361) being the non-cyanogenic ß-glucosidase and Type II clones (eg TRE104) representing the cyanogenic ß-glucosidase (linamarase).   The presence of at least two distinct ß-glucosidase enzymes in white clover leaf tissue has been demonstrated by Maher and Hughes *(9)* and Kakes and Eeltink *(15)* and the non-cyanogenic ß-glucosidase has been purified from young leaves of *lili* clover plants *(16)*.   Physically it is very similar to the cyanogenic ß-glucosidase, having the same subunit molecular weight, the same net charge and being a glycoprotein with N-asparagine linked high mannose oligosaccharides.   In young leaves of *LiLi* plants, it is estimated to be present at about 1/10 the level of the cyanogenic enzyme.

The nucleotide and derived amino acid sequence of the cyanogenic ß-glucosidase from cassava (CAS5) has been published *(12)*.   There is no poly(A) tail on CAS5 but a putative polyadenylation signal is present between bases 1650 and 1662.   The purified native cassava cyanogenic ß-glucosidase behaves as if the N-terminus is blocked and it was not possible to produce an N-terminal peptide sequence.   However, two internal peptide sequences, generated by digestion of the enzyme by a-chymotrypsin are present in CAS5. The identification of these sequences in the derived amino acid sequence of CAS5, confirms it as the cyanogenic ß-glucosidase. Five putative N-asparagine glycosylation sites have been found in the deduced amino acid sequence. Since the N-terminus of the active enzyme is not known, the presence of a signal peptide on the nascent cassava polypeptide cannot be determined, however a hydrophobicity plot of the derived amino acid sequence of CAS5 reveals a very hydrophobic N-terminal 12 amino acid peptide which has a number of features common to signal peptides *(17,18)*. The deduced amino acid sequence of CAS5 has 59% homology with that of TRE104, but only one of the glycosylation sites is conserved (NAT, residues 366-368) and this has an adjacent proline in cassava which may exclude this site from glycosylation.

Recently, 2-deoxy-2-fluoro-ß-D-glucopyranoside has been used to chemically identify a glutamic acid residue as the nucleophilic residue in the active site of a ß-glucosidase from *Agrobacterium (19)*. The sequence which contains this residue, I/V TENG, is present in both of the cyanogenic ß-glucosidases and the non-cyanogenic ß-glucosidase from white clover. The homology of the three plant ß-glucosidases from white clover and cassava and a number of bacterial ß-glucosidases and ß-galactosidases is shown in Figure 8. One of thesignificant differences between the plant and the bacterial enzyme is the presence of a hydrophobic signal sequence at the N-terminus of the plant sequences. It is noticeable that a number of aspartic acid residues are conserved in all these sequences.

Restriction Fragment Segregation Analysis of the *Li* Locus in White Clover

A considerable amount of restriction fragment length polymorphism (RFLP) exists in white clover for sequences with homology to TRE104 and TRE361. The number of restriction fragments varies between plants and co-segregation of cyanogenic ß-glucosidase activity and restriction fragment pattern has been studied in a number of segregating progeny in an attempt to elucidate the genomic organisation of ß-glucosidase genes in this species.

Figure 9 shows the result of crossing a heterozygous *Lili* plant ((Eg x ZD5)83), which was cyanogenic ß-glucosidase positive, with a homozygous *lili* cyanogenic ß-glucosidase null plant ([(Eg x ZD5)83 x ZD5]21). The progeny segregate for presence versus absence of enzyme activity and for the presence or absence of a 5100 base restriction fragment which has homology with the non-cyanogenic clone, TRE361 but not TRE104. The results show that this 5100 base fragment co-segregates with cyanogenic ß-glucosidase activity and must therefore be linked to the *Li* locus *(13)*. The identical cross was analysed with sequences which are unique to the cyanogenic ß-glucosidase clone TRE104 *(14)*. The Southern blot analysis of restriction fragments homologous to this probe also shows co-segregation with enzyme activity. These data are consistent with the proposal that the structural information for both the cyanogenic and the non-cyanogenic ß-glucosidases are linked to the *Li* locus, which controls cyanogenic

CLUSTAL V multiple sequence alignment

```
TRE104      ---------LLSITTTHIHAFKPL-PISFDDFSDLNRSCFAPGFVFGTAS
TRE361      MDFIVAIFALFVISSFTITSTNAVEASTLLDIGNLSRSSFPRGFIFGAGS
CAS5        M--LVLFISLLALTRPAMGTDDDDDNIP----DDFSRKYFPDDFIFGTAT
ASGLUB      MDSPMTDPNTLAAR----------------------FPGDFLFGVAT
BPBGLA      MT--IFQ------------------------------FPQDFMWGTAT
BPBGLB      MSENTFI------------------------------FPATFMWGTST
CSBGLA      MDMS---------------------------------FPKGFLWGAAT
LCPBGAL     MSK---------Q------------------------LPQDFVMGGAT
            *                            ..  *  *  .

TRE104      SAFQYEGAAFEDGKGPSIWDTFTHKYPEKIKDRTNGDVAIDEYHRYKEDI
TRE361      SAYQFEGAVNEGGRGPSIWDTFTHKYPEKIRDGSNADITVDQYHRYKEDV
CAS5        SAYQIEGEATAKGRAPSVWDIFSKETPDRILDGSNGDVAVDFYNRYIQDI
ASGLUB      ASFQIEGSTKADGRKPSIWDAFCNM-PGHVFGRHNGDIACDHYNRWEEDL
BPBGLA      AAYQIEGAYQEDGRGLSIWDTFAHT-PGKVFNGDNGNVACDSYHRYEEDI
BPBGLB      SSYQIEGGTDEGGRTPSIWDTFCQI-PGKVIGGDCGDVACDHFHHFKEDV
CSBGLA      ASYQIEGAWNEDGKGESIWDRFTHQ-KRNILYGHNGDVACDHYHRFEEDV
LCPBGAL     AAYQVEGATKEDGKGRVLWDDFLDK-QGRF----KPDPAADFYHRYDEDL
            .,.* **    *   .**        .    *        *

TRE104      GIMKDMNLDAYRFSISWPRVLPKGKLSGGVNREGINYYNNLINEVLANGM
TRE361      GIMKDQNMDSYRFSISWPRILPKGKLSGGINHEGIKYYNNLINELLANGI
CAS5        KNVKKMGFNAFRMSISWSRVIPSGRRREGVNEEGIQFYNDVINEIISNGL
ASGLUB      DLIKEMGVEAYRFSLAWPRIIPDG--FGPINEKGLDFYDRLVDGCKARGI
BPBGLA      RLMKELGIRTYRFSVSWPRIFPNG--DGEVNQEGLDYYHRVVDLLNDNGI
BPBGLB      QLMKQLGFLHYRFSVAWPRIMPAA---GIINEEGLLFYEHLLDEIELAGL
CSBGLA      SLMKELGLKAYRFSIAWTRIFPDG--FGTVNQKGLEFYDRLINKLVENGI
LCPBGAL     ALAEKYGHQVIRVSIAWSRIFPDG--AGEVEPRGVAFYHKLFADCAAHHI
            * *. *  *. *.    .   . . .*.  *.

TRE104      QPYVTLFHWDVPQALEDEYRGFLGRNIVDDFRDYAELCFKEFGDRVKHWI
TRE361      QPFVTLFHWDLPQVLEDEYGGFLNSGVINDFRDYTDLCFKEFGDRVRYWS
CAS5        EPFVTIFHWDTPQALQDKYGGFLSRDIVYDYLQYADLLFERFGDRVKPWM
ASGLUB      KTYATLYHWDLPLTLMGD-GGWASRSTAHAFQRYAKTVMARLGDRLLAVA
BPBGLA      EPFCTLYHWDLPQALQDA-GGWGNRRTIQAFVQFAETMFREFHGKIQHWL
BPBGLB      IPMLTLYHWDLPQWIEDE-GGWTQRETIQHFKTYASVIMDRFGERINWWN
CSBGLA      EPVVTLHWDLPQKLQDI-GGWANPEIVNYYFDYAMLVINRYKDKVKKWI
LCPBGAL     EPFVTLHHFDTPERLHEA-GDWLSQEMLDDFVAYAKFCFEEFSE-VKYWI
            *. * **    .       ..          .   ..

TRE104      TLNEPWGVSMNAYAYGTFAPGRCSDWLKLNCTGGDSGREPYLAAHYQLLA
TRE361      TLNEPWVFSNSGYALGTNAPGRCSA--SNVAKPGDSGTGPYIVTHNQILA
CAS5        TFNEPSAYVGFAHDDGVFAPGRCSSWVNRQCLAGDSATEPYIVAHNLLLS
ASGLUB      TFNEPWCAVWLSHLYGVHAPG-----------ERN-MEAALAAMHHINLA
BPBGLA      TFNEPWCIAFLSNMLGVHAPG-----------LTN-LQTAIDVGHHLLVA
BPBGLB      TINEPYCASILGYGTGEHAPG-----------HEN-WREAFTAAHHILMC
CSBGLA      TFNEPYCIAFLGYFHGIHAPG-----------IKD-FKVAMDVVHSLMLS
LCPBGAL     TINEPTSMAVQQYTTGTFPPA-----------ESGRFDKTFQAEHNQMVA
            * ***       .   .*.               *
```

Figure 8.   Clustal V multiple sequence alignment *(22)* region of deduced amino acid sequences of eight β-glucosidase genes. TRE104, cyanogenic β-glucosidase, white clover; TRE361, non-cyanogenic β-glucosidase, white clover; CAS5 cyanogenic β-glucosidase, cassava; ASGLUB, cellobiase, *Agrobacterium (23)*; BPBGLA BPBGLB, two β-glucosidases, *Bacillus polymyxa. (24)*; CSBGLA, β-glucosidase, *Caldocellum saccharolyticum (25)*; LCPBGAL, β-phosphogalactosidase, *Lactobaccilus casei (26)*.

Continued on next page

```
TRE104    HAAAARLYKTKYQASQNGIIGITLVSHWF-----EPASKEK-ADVDAAKR
TRE361    HAEAVHVYKTKYQAYQKGKIGITLVSNWL-----MPLDDNSIPDIKAAER
CAS5      HAAAVHQYRKYYQGTQKGKIGITLFTFWY-----EPLSDSKV-DVQAAKT
ASGLUB    HGFGVEASR---HVAPKVPVGLVLNAHSA------IPASDGEADLKAAER
BPBGLA    HGLSVRRFR---ELGTSGQIGIAPNVSWA------VPYSTSEEDKAACAR
BPBGLB    HGIASNLHK---EKGLTGKIGITLNMEHV------DAASERPEDVAAAIR
CSBGLA    HFKVVKAVK---ENNIDVEVGITLNLTPVYLQTERLGYKVSEIEREMVSL
LCPBGAL   HARIVNLYK---SMQLGGQIGIVHALQTVY------PYSDSAVDHHAAEL
            *               .*.

TRE104    GLDFMLGWFMHPLTKGRYPESMRYLV--------RKRLPKFSTEESKELT
TRE361    SLDFQFGLFMEQLTTGDYSKSMRRIV--------KNRLPKFSKFESSLVN
CAS5      ALDFMFGLWMDPMTYGRYPRTMVDLA--------GDKLIGFTDEESQLLR
ASGLUB    AFQFHNGAFFDPVFKGEYPAEMME---ALG-----DRMPVVEAEDLGIIS
BPBGLA    TISLHSDWFLQPIYQGSYPQFLVDWFAEQG-----ATVP-IQDGDMDIIG
BPBGLB    RDGFINRWFAEPLFNGKYPEDMVEWYGTYL-----NGLDFVQPGDMELIQ
CSBGLA    SSQLDNQLFLDPVLKGSYPQKLLDYLVQKDLLDSQKALSMQQEVKENFIF
LCPBGAL   QDALENRLYLDGTLAGEYHQETLA-LVKEILDANHQPMFQSTPQEMKAID
                .    * *                                .

TRE104    GS----FDFLGLNYYSSYYAAKAPRIPNARPAIQTDSLINAT-FEHNGKP
TRE361    GS----FDFIGINYYSSSYISNAPSHGNAKPSYSTNPMTNIS-FEKHGIP
CAS5      GS----YDFVGLQYYTAYYAEPIPPVDPKFRRYKTDSGVNATPYDLNGNL
ASGLUB    QK----LDWWGLN-YYTP-MRV------ADDATPGVEFPATMPAPAVSDV
BPBGLA    EP----IDMIGIN-YYS---MS------VNRFNPEAGFLQSEEIN-MGLF
BPBGLB    QP----GDFLGIN-YYT---RS------IIRSTNDASLLQVEQVH-MEEF
CSBGLA    P------DFLGIN-YYT---RA------VRLYDENSSWIFPIRWEHPAGE
LCPBGAL   EAAH-QLDFVGVNNYFSKWLRAYHGKSETIHNGDGTKGSSVARLQGVGEE
              *  *..  *

TRE104    LGPMAASS----WLCIYP----QGIRKLLLYVKNHYNNPVIY-ITENGR-
TRE361    LGPRAASI----WIYVYPYMFIQEDFEIFCYILKINITILQFSITENGM-
CAS5      IGPQAYSS----WFYIFP----KGIRHFLNYTKDTYNDPVIY-VTENGV-
ASGLUB    K-------TDIGWE-----VYAPALHTLVETLYERYDLP-ECYITENGAC
BPBGLA    V-------TDIGWP-----VESRGLYEVLHYLQK-YGN-IDIYITENGAC
BPBGLB    V-------TDMGWE-----IHPESFYKLLTRIEKDFSKGLPILITENGAA
CSBGLA    Y-------TEMGWE-----VFPQGLFDLLIWIKESYPQ-IPIYITENGAA
LCPBGAL   KLPDGIETTDWDWS-----IYPRGMYDILMRIHNDYPLVPVTYVTENGIG
             *                                    .****

TRE104    --NEFNDPTLSLQESLLDTPRIDYYYRHLYYVLTAIGD-GVNVKGYFAWS
TRE361    --NEFNDATLPVEEALLNTYRIDYYYRHLYYIRSAIRA-GSNVKGFYAWS
CAS5      --DNYNNESQPIEEALQDDFRISYYKKHMWNALGSLKNYGVKLKGYFAWS
ASGLUB    YNMGVE-NGQ-----VNDQPRLDYYAEHLGIVADLIRD-GYPMRGYFAWS
BPBGLA    INDEVV-NGK-----VQDDRRISYMQQHLVQVHRTIHD-GLHVKGYMAWS
BPBGLB    MRDELV-NGQ-----IEDTGRHGYIEEHLKACHRFIEE-GGQLKGYFVWS
CSBGLA    YNDIVTEDGK-----VHDSKRIEYLKQHFEAARKAIEN-GVDLRGYFVWS
LCPBGAL   LKESLPENATP-DTVIEDPKRIDYVKKYLSAMADAIHD-GANVKGYFIWS
            *    *             .     *   ..*.  **

TRE104    LFDNMEWDSGYTVRFGLVL---------------------------
TRE361    FLDCNEWFAGFTVRFGLNFVD-------------------------
CAS5      YLDNFEWNIGYTSRFGLYYVDYKNNLTRYPKKSAHWFTKFLNISVNANNI
ASGLUB    LMDNFEWAEGYRMRFGLVHVDYQTQ-VRTVKNSGKWYSALASGFPKGNH-
BPBGLA    LLDNFEWAEGYNMRFGMIHVDFRTQ-VRTPKESYYWYRNVVS----NNW-
BPBGLB    FLDNFEWAWGYSKRFGIVHINYETQ-ERTPKQSALWFKQMMA-----KNG-
CSBGLA    LMDNFEWAMGYTKRFGIIYVDYETQ-KRIKKDSFYFYQQ----YIKEN--
LCPBGAL   LQDQFSWTNGYSKRYGLFFVDFPTQ-NRYIKQSAEWFKSVSETHI-----
            *   *   *.    *.*.
```

Figure 8. Continued.

1 2 3 4 5   6 7 8 9 10

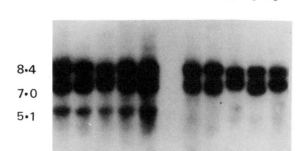

Figure 9. Restriction fragment length polymorphism segregation analysis of the progeny from the cross (Eg x ZD5)83 *Li li* x [(Eg x ZD5)83 x ZD5]21 *li li*. Wells 1-5 cyanogenic β-glucosidase-positive *Li li* plants; Wells 6-10 cyanogenic β-glucosidase-negative *li li* plants.   Scale in Kbases. Complete pTRE36 insert used as a probe. (Reproduced with permission from ref. 13).

ß-glucosidase activity in white clover.   There is also evidence for a third ß-glucosidase gene in white clover since segregation analysis of a different cross ([(Eg xD4) x ZD5]3 *Lili* x ZD5 *lili*) indicates that one restriction fragment with homology to TRE361 segregates independently of *Li* alleles in the resulting progeny *(13)*.

Localisation of Cyanogenic ß-glucosidase Gene Expression in Cassava

The distribution of cyanogenic ß-glucosidase activity differs in white clover and cassava, since cassava roots contain the enzyme and white clover roots do not. The non-isotopic digoxigenin labelling system was used to visualise the presence of cyanogenic ß-glucosidase mRNA in cells of young leaves of cassava *(20)*. Strong hybridisation to antisense riboprobes produced from the cDNA clone CAS5 indicated localisation of cyanogenic ß-glucosidase gene expression to laticifers (latex vessels) and to no other cells of the leaf. This conclusion was supported by the demonstration of cyanogenic ß-glucosidase mRNA in exuded latex (Figure 10). *Manihot* and *Hevea* are both members of the Euphorbiaceae and both are species which have multicelled branching latex vessels. Although studies on the development, structure and biochemistry of latex vessels in cassava are very limited, an extensive literature exists for these cells in *Hevea braziliensis* L (rubber). The latex is under pressure and oozes from wounded surfaces where it coagulates.   It is widely accepted that latex has a protective or defensive function and enzymes with antifungal activity, such as chitinases, have been found in *Hevea* latex *(21)*.

The presence of cyanogenic ß-glucosidase in cassava latex was further investigated in an experiment in which enzyme activity was assayed in latex

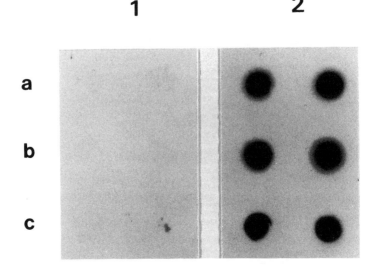

Figure 10. Dot-blot of cassava latex probed with sense and antisense riboprobes of cyanogenic β-glucosidase cDNA clone (pCAS5). (1) sense probe; (2) antisense probe; (a) leaflet mid vein; (b) petiole; (c) stem 2.5 μl latex applied directly onto Hybond N membrane following denaturation of RNA in formamide-formaldehyde MOPS buffer.

recovered from syringe needle wound sites on petioles of four leaves on a single plant following the removal of different numbers of leaflets. This analysis shows that levels of enzyme activity in the petiole latex are related to the number of leaflets attached, with the lowest levels present in petioles with all leaflets removed (20). These results suggest that the cyanogenic ß-glucosidase in cassava is synthesised in the latex vessels of young leaves, and transported to the latex vessels of the petiole and possibly to the vessels of the stem and roots.

Literature Cited

1.   Collinge, D.B.; Hughes, M.A. *Arch. Biochem. Biophys*. **1982** 218: 38-45.
2.   Koch, B.; Neilsen, V.S.; Halkier, B.A.; Olsen, C.E.; Moller, B.L. *Arch.Biochem. Biophys*. **1992**, 292 : 141-150.
3.   Fan, T,W-M.; Conn, E.E. *Arch. Biochem. Biophys*. **1988**, 243 : 361-373.
4.   Frehner, M.; Conn, E.E. *Plant Physiol*. **1987**, 84 : 1296-1300.
5.   Selmar, D.; Lieberei, R.; Biehl, B.; Voight, J. *Plant Physiol*. **1987**, 83 : 557-563.
6.   Abrol, Y.P.; Conn, E.E. *Phytochemistry* **1966,** 5 : 237-242.

7.   Armstrong, H.E.; Armstrong, E.F.; Horton, E. *Roy. Soc. (Lond.) Proc. B*, **1913**, 86 : 262-269.

8.   Corkill, L. *NZ J. Sci. Technol. B* **1942**, 23 : 178-193.

9.   Maher, E.P.; Hughes, M.A. *Biochem. Genet.* **1973**, 8 : 13-26.

10.   Hughes, M.A.; Dunn, M.A.; Pearson, J.R. *Heredity*, **1985**, 55 : 387-391.

11.   Dunn, M.A.; Hughes, M.A.; Sharif, A.L. *Arch. Biochem. Biophys.* **1988**, 260 : 561-568.

12.   Hughes, M.A.; Brown, K.; Pancoro, A.; Murray, S.B.; Oxtoby, E.; Hughes, J. *Arch. Biochem. Biophys.* **1992**, 295 : 273-279.

13.   Hughes, M.A.; Sharif, A.L.; Dunn, M.A.; Oxtoby, E.; Pancoro, A. *Plant Mol. Biol.* **1990**, 14 : 407-414.

14.   Oxtoby, E.; Dunn, M.A.; Pancoro, A.; Hughes, M.A. *Plant Mol. Biol.* **1991**, 17 : 209-219.

15.   Kakes, P.; Eeltink, H. *Z. Naturforsch. Sect. C. Biosci.* **1985**, 40 : 509-513.

16.   Hughes, M.A.; Dunn, M.A. *Plant Mol. Biol.* **1982**, 1 : 169-191.

17.   Watson. M.E.E. *Nucleic Acids Res.* **1984**, 12 : 5145-5164.

18.   von Heijne. *In Plant Molecular Biology* **1991**, 2, Herrmann, R.G.; Larkins, B., Eds., Plenum Press, N.Y., USA. 1991 : 583-593.

19.   Withers, S.G.; Warren, R.A.J.; Street, I.P.; Rupitz, K.; Kempton, J.B.; Aebersold, R. *J. Am. Chem. Soc.* **1990**, 112 : 5887-5889.

20.   Pancoro, A.; Hughes, M.A. *The Plant Journal:* **1992**, 2 : 821-827.

21.   Martin, N.N. *Plant Physiol.* **1990**, 95 : 469-476.

22.   Higgins, D.G.; Sharp, P.M. *Gene* **1988**, 73 : 237-244.

23.   Wakarchuk, W.W.; Greenberg, N.M.; Kilburn, D.G.; Miller, R.C. Jr; Warren, R.A.J. *J. Bacteriol*, **1988**, 170 : 301-307.

24.   Gonzales-Candelas, L.; Ramon, D.; Polaina, J. *Gene* **1990**, 95 : 31-38.

25.   Love, D.R.; Bergquist, P.L. *Mol. Gen. Genet.* **1988**, 213 : 84-92.

26.   Porter, E.V.; Chassy, B.M. *Gene* **1988**, 62 : 263-276.

RECEIVED February 1, 1993

Chapter 12

Enzymology of Cyanogenesis in Rosaceous Stone Fruits

Jonathan E. Poulton

Department of Biological Sciences, University of Iowa,
Iowa City, IA 52242

Mature black cherry (*Prunus serotina*) seeds accumulate the
cyanogenic diglycoside (*R*)-amygdalin (the β-gentiobioside of
(*R*)-mandelonitrile). Upon tissue disruption, amygdalin is rapidly
catabolized to HCN and benzaldehyde by the enzymes amygdalin
hydrolase, prunasin hydrolase and mandelonitrile lyase. These
glycoproteins were purified to homogeneity and their major kinetic
and molecular properties characterized. Aspects of the temporal
and spatial regulation of cyanogenesis in maturing cherry fruits
were investigated using monospecific polyclonal antisera raised
against each of the deglycosylated proteins. The three catabolic
enzymes, which first appeared within developing seeds about six
weeks after flowering, were localized at the tissue and subcellular
levels by colloidal gold immunocytochemistry. Amygdalin hydrolase
and prunasin hydrolase were found specifically within protein
bodies of the procambium, while mandelonitrile lyase was primarily
located within protein bodies of the cotyledonary parenchyma cells
and with lesser amounts within the procambium. Amygdalin
localization, which would reveal how premature cyanogenesis is
avoided in undamaged seeds, is under investigation.

Many important food plants, including sorghum, cassava, lima beans and cherries,
accumulate cyanogenic glycosides, which upon hydrolysis liberate the toxic
respiratory poison hydrogen cyanide (HCN) (*1, 2*). This phenomenon of
cyanogenesis accounts for numerous cases of acute and chronic cyanide poisoning
of animals including man (*3*). A thorough understanding of the biochemistry and
physiology of cyanogenesis is highly desirable to facilitate attempts to reduce the
toxicity of such plants.

0097–6156/93/0533–0170$06.25/0
© 1993 American Chemical Society

Historical Introduction

The kernels of *Prunus* species (Rosaceae) are a rich source of the cyanogenic diglucoside (R)-amygdalin and of the complex enzyme system commonly known as emulsin, which attacks a wide variety of α- and β-glycosidic bonds. Since 1837 when Liebig and Wöhler (4) first described the action of emulsin on amygdalin, several attempts have been made to determine how many β-glycosidases are involved in amygdalin catabolism. Supporting earlier work by Armstrong *et al.* (5), Haisman and Knight (6) concluded from kinetic data that three enzymes, namely amygdalin hydrolase (AH), prunasin hydrolase (PH) and mandelonitrile lyase (MDL), catalyzed the degradation of amygdalin to HCN and benzaldehyde in almonds (Figure 1). The two β-glucosidases, which appeared specific for their respective cyanogenic substrates, were later identified by continuous electrophoresis in a free buffer film (7). Because these enzymes were not completely resolved and purified to homogeneity however, description of their kinetic and molecular properties was not possible. In 1974, two β-glucosidase isozymes were isolated from sweet almonds by Lalegerie (8). These enzymes differed from those described by Haisman and Knight in that each was active toward both amygdalin and and its respective monoglucoside prunasin and they displayed 4-5 fold higher K_m values. In other laboratories, almond emulsin was found to contain three (9) or at least four (10) β-glucosidase isozymes. Unfortunately, the relationship of these hydrolases to AH and PH remains unknown because synthetic rather than natural substrates were used to assay enzyme activity. Thus, although it appears likely that amygdalin hydrolysis in almonds occurs in a stepwise fashion with prunasin acting as intermediate, there is little concensus as to the number and nature of β-glycosidases involved.

In view of the difficulties in obtaining fresh almonds locally and the uncertainties of utilizing almond emulsin as a source of β-glycosidases, we selected black cherry (*Prunus serotina*) to reinvestigate the enzymology of amygdalin catabolism. The ready availability of this species within the Midwest allows sampling of tissues as necessary throughout the growing season. As in other *Prunus* species (5, 11, 12), its young leaves and immature fruits contain prunasin, but amygdalin predominates in mature fruits. Using amygdalin and prunasin as substrates during enzyme purification, our early studies with *P. serotina* seed homogenates showed that amygdalin is hydrolyzed to mandelonitrile by two distinct β-glycosidases (AH and PH) which were easily resolvable by ion-exchange chromatography (13). The subsequent dissociation of mandelonitrile to HCN and benzaldehyde is catalyzed by the flavoprotein MDL, which constitutes approx. 5-10% of the soluble proteins of black cherry seeds. As described below, *P. serotina* AH, PH and MDL have now been purified to homogeneity, allowing characterization of their major kinetic and molecular properties. Monospecific polyclonal antibodies were generated against each protein for use in immunocytochemical localization studies and in screening cDNA expression libraries.

Purification and Characterization of *P. serotina* AH, PH and MDL

Four isozymes of AH (designated AH I, I′, II, II′) were purified to virtual homogeneity from seed homogenates by conventional protein purification techniques (*14*). All isozymes are monomeric glycoproteins with native molecular masses of approx. 52 kD. They show similar kinetic properties (K_m, V_{max}, pH optimum of 4.5-5.5) but differ in their isoelectric points and N-terminal amino acid sequences (Table I) (*15*). Their glycoprotein character, first suggested by Con A-Sepharose 4B chromatography, was supported by positive PAS-staining on PAGE and IEF gels.   Upon deglycosylation with TFMS (*16*), the subunit molecular mass of these isozymes (as observed by SDS-PAGE) fell from 62 kD to 57 kD (*15*). These isozymes are extremely stable for several months at 4°C in the presence of 0.4% (w/v) sodium azide.

The substrate specificities of AH I and II were investigated in detail (*14*). Among the naturally occurring glycosides tested, the endogenous cyanogenic diglucoside (R)-amygdalin was most rapidly hydrolyzed (Table II). AH I and II had K_m values for amygdalin of 2.5 and 2.1 mM, respectively, with corresponding V_{max} values of 41.74 and 43.86 mmol glucose/h/mg protein. The aglycone specificity of AH I and II was demonstrated by their much lower activity toward the aliphatic cyanogenic diglucosides linustatin and neolinustatin. AH I and II lacked activity toward β-gentiobiose and did not hydrolyze the cyanogenic monoglucosides (R)-prunasin, (S)-sambunigrin and linamarin. Both isozymes were highly active toward the fluorogenic substrate 4-methylumbelliferyl-β-D-glucoside (4-MUGlc).   Additionally, both AH I and II hydrolyzed several *p*- and *o*-nitrophenyl-β-D-glycosides.   These synthetic compounds provided valuable information concerning how the isoforms respond to variations in the nature and linkage of the sugar moiety within the glycosidic substrate.   Whereas *p*-nitrophenyl-β-D-glucoside (PNPGlc) was hydrolyzed efficiently, little or no activity was observed with the corresponding β-linked galactoside, xyloside, mannoside, or N-acetylglucosaminide. Mixed substrate experiments indicate that amygdalin, 4-MUGlc and PNPGlc were hydrolyzed at the same active site. Both AH I and II require a β-linkage in their substrates as shown by the failure to hydrolyze *p*-nitrophenyl-α-D-glucoside and 4-methylumbelliferyl-α-D-glucoside.

Three isoforms of prunasin hydrolase (designated PH I, IIa and IIb) were purified to apparent homogeneity by conventional purification techniques (*17*). PH I and PH IIb are monomeric (68 kD) whereas PH IIa is dimeric (140 kD). All are glycoproteins based on their binding to Con A-Sepharose 4B with subsequent elution by α-methyl-D-glucoside. After deglycosylation of PH by TFMS, they are prone to aggregation (*15*). When presented several potential glycosidic substrates at the optimum pH (pH 5.0), these isoforms exhibit a narrow specificity towards (R)-prunasin (Table III). By contrast, they are inactive towards the diglucosides (R)-amygdalin, linustatin and neolinustatin and the monoglucosides (S)-dhurrin and linamarin. K_m values for (R)-prunasin for PH I, IIa, and IIb are 1.6, 2.3 and 1.4 mM, respectively. PH I and IIb possess 5-fold greater V_{max}/K_m values than PH IIa. Like AH, the PH isoforms efficiently utilized PNPGlc and *o*-nitrophenyl-β-D-glucoside (ONPGlc) as substrates but

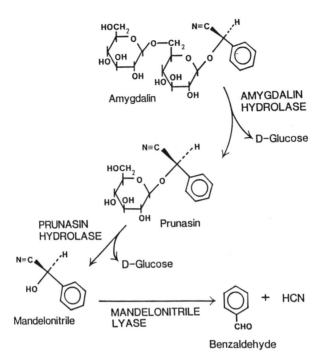

Figure 1. Catabolic pathway for (R)-amygdalin by almond emulsin and homogenates of mature *Prunus serotina* seeds.

Table I. Comparison of the N-Terminal Amino Acid Sequences of
Black Cherry AH and PH Isozymes

AH I	A	K	T	D	P	P	I	H	C	A	S	L	X	R	S	S	–
AH I′	A	K	T	D	P	P	I	H	F	A	S	L	X	R	S	–	
AH II		A	D	P	P	I	H	I	A	S	L	X	–				
AH II′		T	D	P	P	I	H	I	A	S	L	X	R	S	X	–	
PH I	X	X	T	Y	P	P	V	V	X	A	T	L	X	R	T	H	–
PH IIa	A	G	T	Y	P	P	V	V	X	A	T	L	X	R	T	H	–
PH IIb	X	G	T	Y	P	P	V	V	L	A	T	L	X	R	T	H	–

NOTE: Amino acids are indicated by single-letter IUPAC nomenclature with the letter X denoting position of unidentified residue. Sequences have been aligned to maximize similarity. (Adapted from ref. 15.)

Table II. Substrate Specificity of Amygdalin Hydrolase I and II

Substrate	Relative activity	
	AH I	AH II
(R)-Amygdalin	100	100
o-Nitrophenyl-β-D-glucoside	84	76
p-Nitrophenyl-β-D-glucoside	68	74
p-Nitrophenyl-β-D-galactoside	16	9
p-Nitrophenyl-β-D-xyloside	0	0
p-Nitrophenyl-N-acetyl-β-D-glucosaminide	0	0
p-Nitrophenyl-β-D-mannoside	0	0
p-Nitrophenyl-α-D-mannoside	0	0
4-Me-umbelliferyl-β-D-glucoside	110	168
4-Me-umbelliferyl-α-D-glucoside	0	0
Neolinustatin	27	34
Linustatin	10	14
Neocycasin A	11	12
Cycasin	6	8
Salicin	9	8
(R)-Prunasin	0	0
(S)-Sambunigrin	2	1

NOTE: β-Glycosidase activity was assayed as previously described (14). Activities are expressed as a percentage of the rate of amygdalin hydrolysis. AH I and II were inactive towards the following substrates: (S)-dhurrin, β-gentiobiose, linamarin, laminarin, cellobiose, sucrose, lactose, maltose, methyl-β-D-glucoside and phenyl-β-D-glucoside.

Table III. Substrate Specificity of Prunasin Hydrolase I, IIa and IIb

Substrate	Relative activity		
	PH I	PH IIa	PH IIb
(R)-Prunasin	100	100	100
ortho-Nitrophenyl-β-D-glucoside	119	226	105
para-Nitrophenyl-β-D-glucoside	49	63	43
para-Nitrophenyl-β-D-galactoside	15	24	16
para-Nitrophenyl-β-D-xyloside	0	1	0
para-Nitrophenyl-α-D-glucoside	0	0	0
para-Nitrophenyl-N-acetyl-β-D-glucosaminide	0	0	0
para-Nitrophenyl-β-D-mannoside	0	0	0
para-Nitrophenyl-α-D-mannoside	0	0	0
(S)-Sambunigrin	13	14	10
Salicin	8	6	5
Neocycasin A	6	0	7
4-Methylumbelliferyl-β-D-glucoside	3	14	5
Cycasin	2	3	0
(R)-Amygdalin	0	0	0
(S)-Dhurrin	0	0	0
Linamarin	0	0	0
Linustatin	0	0	0
Neolinustatin	0	0	0

NOTE: β-Glycosidase activity was assayed as previously described (*17*). Activities are expressed as a percentage of the rate of prunasin hydrolysis. PH I, IIa and IIb were inactive towards the following substrates: β-gentiobiose, laminarin, cellobiose, sucrose, lactose, maltose, methyl-β-D-glucoside and phenyl-β-D-glucoside.

showed little or no activity toward other p-nitrophenyl-β-D-glycosides tested. Likewise, p-nitrophenyl-α-D-glucoside was not accepted as substrate indicating a requirement for a β-glucosidic linkage. As shown by mixed substrate studies, prunasin, PNPGlc and ONPGlc were hydrolyzed at the same active site.

Further characterization studies showed that cations are apparently not required for AH or PH activities. Addition of Cu^{2+}, Mg^{2+}, Zn^{2+}, Pb^{2+}, Fe^{2+}, Fe^{3+} and Ag^+ ions at 0.1 mM concentration had little or no effect (14, 17). Likewise, chelators, thiol reagents and sulfhydryl reagents were without significant effect on enzyme activity. In contrast, both enzymes were inhibited by the pyrrolizidine alkaloid castanospermine, and AH isoforms were competitively inhibited by the reaction product prunasin (K_i, 0.68 mM).

Sequencing the N-termini of PH isoforms indicated no unequivocal differences between them nor did it shed light upon the physical relationship which might exist between PH IIa and the monomeric PH isoforms (Table I). Interestingly however, extensive homology (53-75%) existed in primary structure between PH and AH isozymes, including the Pro-Pro dipeptide at residues 5 and 6. Immunological data also pointed to structural similarity between AH and PH. Antisera raised against either deglycosylated enzyme cross-reacted with the other hydrolase, suggesting that they share common epitopes (15). Whether this similarity extends to the remaining coding regions, including potential N-glycosylation sites, will be revealed by sequencing their cDNA clones. These efforts will also furnish information about their respective signal peptide sequences, their 5´- and 3´-untranslated nucleotide regions, and their possible synthesis as preproproteins.

Five multiple forms (two major, three minor) of MDL were purified to virtual homogeneity from mature black cherry seeds by Con A-Sepharose 4B chromatography and chromatofocusing (18). These forms are monomers which differ only slightly with regard to molecular mass (57-59 kD) and isoelectric point (pI 4.58-4.63). Because individual P. serotina MDL isozymes are difficult to fully resolve, an isozyme mixture was submitted for N-terminal sequencing (19). The N-terminus, apparently common to all isozymes submitted, was L-A-T-T-S-N-H-D-F-S-Y-L-F-R-A-Y. This is completely identical with the N-terminus of P. lyonii MDL (courtesy of Prof. Eric E. Conn) except at positions 6 and 13; both might represent single base changes. Data obtained with Con A-Sepharose 4B chromatography and absorption spectroscopy strongly suggest that all forms are glycoproteins which contain FAD as prosthetic group. Upon chemical deglycosylation by TFMS, the molecular mass was reduced to 50.7 kD (19). Two-dimensional gel analysis of deglycosylated MDL revealed the presence of several subunit isoforms of similar molecular mass but differing slightly in isoelectric point. Detailed comparative kinetic studies of the two major forms (Forms 4 and 5) were conducted (18). The pH optima for Forms 4 and 5 lie between 6.0 and 7.0, and both proteins retain similar activities upon storage at 4°C and -20°C. K_m values for (R,S)-mandelonitrile of 0.172 and 0.174 mM were determined for Forms 4 and 5, respectively. Neither lyase form exhibits a metal ion requirement, and iodoacetic acid inhibits these proteins to equal extents while other sulfhydryl reagents were ineffective. Both forms are inhibited by unsubstituted and *o*- and *p*-hydroxy-substituted aromatic substrate analogs and by hexanoic acid.

One of the complexities of the *Prunus* cyanogenic system yet to be satisfactorily unraveled is the nature and physiological significance of the microheterogeneity shown by AH, PH and MDL (4, 3 and 5 isoforms, respectively) (*15, 17, 18*). Such multiplicity, which is also shown by analogous enzymes from other rosaceous stone fruits (*10, 20*), might well result from allelic variance because, as enzyme source, we use seeds collected from many individual trees or, at best, a single tree. Alternative potential sources of heterogeneity include the existence of multigene families, differences in post-translational modifications, and, in the case of the dimeric PH IIa isoform, aggregation-dissociation phenomena (*21-23*). In our opinion, partial proteolysis during extraction and purification of AH, PH and MDL does not contribute significantly to their observed multiplicity because inclusion of several protease inhibitors did not reduce their heterogeneity (*14, 17, 18*). Furthermore, the minor but significant differences seen in the N-termini of the four AH isozymes (Table I) cannot be readily explained by proteolysis. At least for MDL, removal of carbohydrate side-chains by TFMS did not reduce apparent heterogeneity. Deglycosylated MDL still displayed 4-6 discrete polypeptide spots upon 2D-SDS-PAGE (*19*).

Preliminary evidence favoring allelic variance as a source of AH microheterogeneity was obtained last year by screening isozyme patterns of individual seeds (*15*). The predominant AH isozyme(s) clearly varied from seed to seed, indeed even between seeds from the same tree (Figure 2). This finding explains, at least in part, our observation that the relative levels of the four AH isozymes differ between purification runs. It is possible that some of the heterogeneity shown by PH and MDL may have similar origins, but at present we cannot rule out the possibility of different post-translational modifications or the existence of multigene families. It should be noted that polymorphism exists for β-glycosidases from other species. With over 30 different alleles, maize β-glucosidase is the most polymorphic enzyme locus known in any organism (*24*).

Temporal and Spatial Expression of AH, PH and MDL in Maturing Fruits

The biochemistry of cyanogenic glycoside accumulation by maturing fruits and seeds has so far received little attention. Without exception, where cyanoglycoside levels have been monitored during fruit development, no attempts were made to correlate them with changes in levels of enzymes involved in cyanoglycoside metabolism (*11, 12, 25-27*). As a result, it has generally remained unclear exactly when fruits first develop their capacity for cyanogenesis and furthermore, whether the cyanoglycosides found in fruits and seeds are actually synthesized *in situ* or are translocated from sites of synthesis outside the fruit (*27*). One major goal of our research has therefore been to learn how cyanophoric fruits, such as those of *Prunus*, develop their capacity for cyanogenesis during ripening. Using maturing black cherry fruits, our aim is to correlate the biochemistry of glycoside accumulation with the anatomical changes that occur during fruit maturation. By concomitantly assaying AH, PH and MDL levels, changes in the capacity of developing fruits to exhibit cyanogenesis could also be assessed.

The development of *P. serotina* fruits shows three distinct growth phases (Figure 3). During Phase I (0-28 days after flowering (DAF)), the fruits exhibit a dramatic increase both in size and in fresh and dry weights. During mid-Phase I, the endocarp begins differentiation and, by 28 DAF, is quite woody and encloses glassy-looking nucellar and endosperm tissues. Thereafter, a period of retarded growth (Phase II; 29-65 DAF) ensues during which the embryo develops within and at the expense of the nucellus and endosperm. Cotyledons become visible to the naked eye approx. 42 DAF and, by the end of Phase II (65 DAF), they occupy most of the seed volume, while the residual endosperm tissue is restricted to two thin strips of cells associated with the surrounding testa. The last phase of fruit maturation (Phase III; 66-81 DAF) is characterized by a marked increase in fresh and dry weights as the mesocarp ripens and the cherries turn black in color.

During the 1990 season, biochemical changes related to cyanogenesis were monitored during fruit maturation (*29*). At weekly intervals from flowering until maturity, whole fruits were analyzed for: (1) fresh and dry weights, (2) prunasin and amygdalin levels, and (3) levels of AH, PH and MDL (Figures 3-6). After endocarp development began (28 DAF), fruits were also dissected into two fractions, namely the developing seeds (the contents of the endocarp) and the remaining tissues (the pericarp) for separate analysis. During Phase I, immature fruits accumulated prunasin (mean: 3 μmol/fruit) but were acyanogenic since they lacked the catabolic enzymes. The ability of developing fruits to release significant levels of HCN upon tissue disruption was first noted during mid-Phase II. Coincident with cotyledon development during mid-Phase II, the seeds began accumulating amygdalin (mean: 3 μmol/seed) while prunasin levels rapidly declined and were negligible by maturity. At this time, the levels of AH, PH and MDL increased rapidly and specifically within seeds (Figure 5). Such dramatic increases in enzyme activity are probably the result of *de novo* synthesis because immunoblotting data (obtained by using affinity purified monospecific polyclonal antibodies (*15, 19*)) revealed that these proteins were absent from developing seeds until this time (Figure 6). During Phases II and III, the pericarp also accumulated amygdalin whereas its prunasin content declined towards maturity. Lacking the catabolic enzymes, the pericarp remained acyanogenic throughout all developmental stages.

The Role of Compartmentation in Preventing Premature Cyanogenesis in Undamaged *P. serotina* Seeds

Because tissue disruption is required for large-scale HCN release by cyanogenic plants, it seems probable that the cyanogens and their respective catabolic enzymes have different tissue or subcellular localizations in undamaged organs (*1, 30*). Alternatively, they might co-exist within the same subcellular compartment if the activity of the hydrolases were suppressed by endogenous inhibitors (*31*).

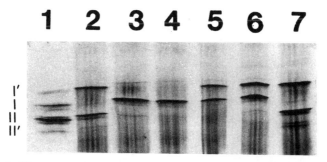

Figure 2. Heterogeneity observed in AH isozyme patterns of individual seeds. AH was extracted from individual seeds and partially purified by Con A-Sepharose 4B chromatography before being submitted to IEF on Phastgel IEF 3/9 gels followed by silver staining (*15*). Lane 1, highly purified AH, exhibiting all four isozymes; lanes 2 & 3, lanes 4 & 5, and lanes 6 & 7 show isozyme patterns from pairs of seeds taken from three different trees located at least 2 miles apart. (Adapted from ref. 15.).

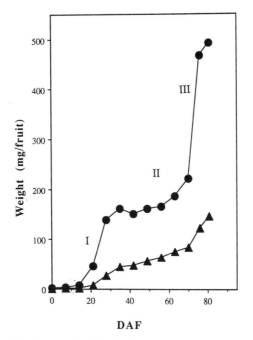

Figure 3. Growth of maturing black cherry fruits from flowering to maturity. Mean fresh weights (●) and dry weights (▲) were determined as previously described (*29*). (Reproduced with permission from ref. 29. Copyright 1992 American Society of Plant Physiologists).

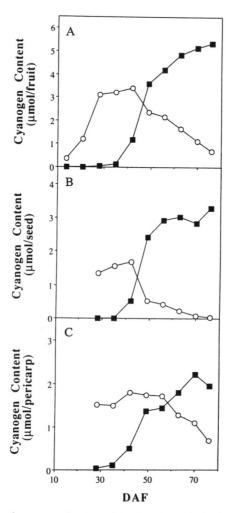

Figure 4. Levels of cyanogenic glycosides in whole fruits (A), seeds (B), and pericarps (C) during fruit maturation. Prunasin (O) and amygdalin (■) were extracted from pulverized tissues (n = 11-100), depending on developmental stage) in boiling 70% methanol, resolved by HPLC, and quantitated after enzymic hydrolysis (29). Each data point represents the mean of two replicates whose variability did not exceed the dimensions of the symbol shown. (Reproduced with permission from ref. 29. Copyright 1992 American Society of Plant Physiologists).

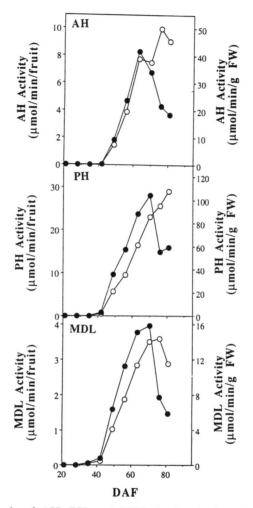

Figure 5. Levels of AH, PH and MDL in developing *P. serotina* seeds. Extracts of developing seeds (*n* = 8-750, depending on developmental stage) were partially purified as previously described (*29*). Enzyme activities are expressed on a per fruit (O) or on a per g fresh weight (●) basis. Each data point represents the mean of three replicates whose variability did not exceed the dimensions of the symbol shown. (Reproduced with permission from ref. 29. Copyright 1992 American Society of Plant Physiologists.)

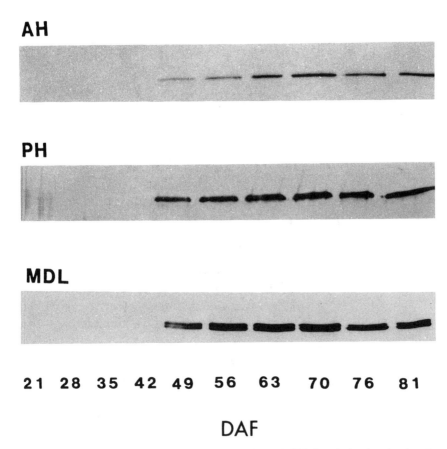

AH

PH

MDL

21 28 35 42 49 56 63 70 76 81

DAF

Figure 6.   Immunoblot analysis of AH, PH and MDL levels in developing *P. serotina* seeds.   Equal volumes (30 μL) of crude seed homogenates, derived from maturing seeds as previously described (*29*), were subjected to SDS-PAGE and immunoblot analysis using polyclonal antibodies monospecific for each protein. (Reproduced with permission from ref. 29. Copyright 1992 American Society of Plant Physiologists).

In this case, cyanogenesis would procede upon tissue damage if these inhibitors were by some means diluted out, degraded, or released from complexation with the hydrolases.

Recent studies involving a wide range of plant species and organs have not yet revealed a common compartmentation strategy for avoiding premature cyanogenesis (*30*). In leaves of 6-day-old light-grown sorghum seedlings, we showed that compartmentation occurs at the tissue level (*32*). The cyanoglucoside (*S*)-dhurrin is located in vacuoles of epidermal cells, whereas its β-glucosidase and α-hydroxynitrile lyase exist almost exclusively in the underlying mesophyll cells (*32, 33*). These enzymes are located in the chloroplasts and cytosol, respectively (*34*). It therefore seems likely that the large-scale hydrolysis of dhurrin occurs only after extensive disruption of the plant tissue. For other cyanogenic species (*35-38*), localization data have largely been interpreted in terms of compartmentation at the subcellular level, although in most cases both components of the "cyanide bomb" have not been unequivocally localized within the same cells. For example, while linamarin and linamarase show similar radial gradients in tissue blocks cut from cassava root cross-sections (36), this observation does not of itself preclude tissue level compartmentation because the exact locations of both glucoside and glucosidase within the tissue remain unknown.

To understand how premature cyanogenesis is avoided in undamaged, mature black cherry seeds requires the unequivocal localization of amygdalin and its catabolic enzymes. At maturity, more than 95% of the seed volume is composed of two fleshy cotyledons enclasping a small embryonic axis. A complex procambial network, surrounded by a single layer of bundle sheath cells, ramifies throughout the cotyledonary storage parenchyma (Figure 7). Lying external to the embryo are two thin strips of residual endosperm tissue, which are associated with the surrounding testa. The tissue and subcellular localizations of AH, PH and MDL within seeds were determined by colloidal gold immunocytochemistry (*39*). All three enzymes reside within protein bodies (Figures 8 & 9), but they show very different tissue-specific localizations. Silver-enhancement colloidal gold immuno–cytochemistry clearly established that AH and PH are confined to the procambium and are absent from the cotyledonary storage parenchyma, bundle sheath and endosperm (data not shown). By contrast, highest levels of MDL were observed in protein bodies of the cotyledonary parenchyma cells. This location supports the notion that MDL, which constitutes 5-10% of the seed's soluble proteins, might serve an additional role as a storage protein. Lesser amounts of MDL were observed throughout the procambium, but this enzyme, like AH and PH, appears absent from the bundle sheath and endosperm. A further level of organizational complexity was also detected by immunocytochemistry, namely cell-specific distribution of AH and PH within the procambium. Whereas AH occurred throughout the procambium, PH was restricted to peripheral cells of this meristematic tissue (*39*).

With localization of AH, PH and MDL now complete, we are focusing our attention upon localizing amygdalin. Our early experiments showed that groups of cells carefully excised from cherry cotyledons contained amygdalin,·but the exact nature of these excised cells could not be established. The absence of AH

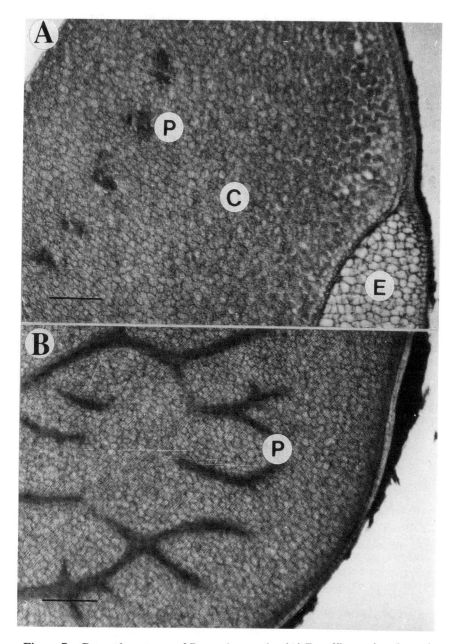

Figure 7.   General anatomy of *P. serotina* seeds. (A) Paraffin section through cotyledon (C) and endosperm (E) tissues showing procambium (P) in transverse view. (B) Section through cotyledon showing procambium (P) in longitudinal view. Scale bar equals 200 μm. (Reproduced with permission from ref. 39. Copyright 1992 American Society of Plant Physiologists).

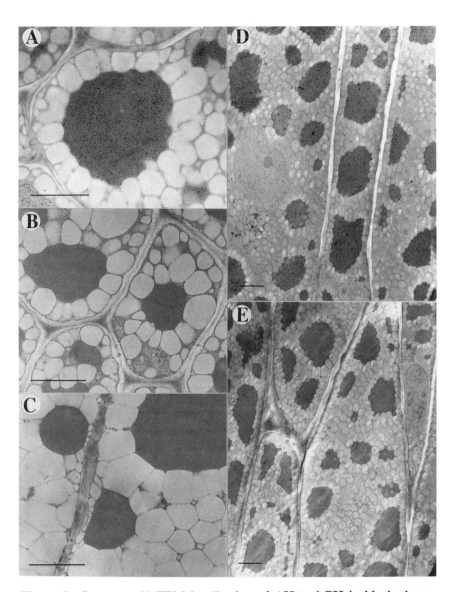

Figure 8. Immunogold TEM localization of AH and PH in black cherry seeds. Thin sections of procambial (A, B, D and E) or cotyledonary parenchyma cells (C) were exposed to the following primary antisera: (A and C) affinity-purified rabbit anti-AH; (D) affinity-purified rabbit anti–PH; (B and E) preimmune serum. Bound antibodies were visualized using colloidal gold-conjugated GAR IgG. The elongated nature of the procambial cells is illustrated by the transverse (A and B) and longitudinal (D and E) sections of these cells. Scale bar equals 2 μm. (Reproduced with permission from ref. 39. Copyright 1992 American Society of Plant Physiologists).

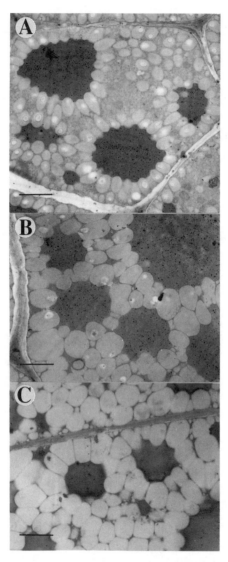

Figure 9.   Immunogold TEM localization of MDL in black cherry seeds.
Procambial (A) and cotyledonary parenchyma (B) cells were treated with
anti-MDL antiserum.   Bound MDL antibodies were localized by GAR
immunogold IgG (39).   As a control, cotyledonary parenchyma cells (C) were
also exposed to preimmune serum before treatment with immunogold IgG.
Scale bar equals 2 μm.   (Reproduced with permission from ref. 39.
Copyright 1992 American Society of Plant Physiologists).

and PH in bundle sheath and cotyledonary parenchyma cells renders these tissues particularly attractive potential locations for amygdalin. However, the coexistence of amygdalin with AH and PH in the procambium (perhaps in different subcellular compartments) cannot be ruled out yet. In common with other unpigmented, water-soluble natural products of low molecular weight, the cyanogenic glycosides are notoriously difficult to localize at the tissue and subcellular levels (e.g. *40, 41*). During the past year, we approached the problem of amygdalin localization through protoplast isolation and subfractionation. We successfully developed techniques to obtain protoplasts in moderate yields from black cherry seeds by cellulase/pectinase digestion. However, although our protoplast populations appear quite uniform, they are undoubtedly derived from several tissues (e.g. storage parenchyma, bundle sheath and procambium). We therefore feel that this approach will at best prove ambiguous in localizing amygdalin. Given the complexity of the tissue- and cell-specific localizations exhibited by AH, PH and MDL, we urgently require alternative approaches having sufficient definition and specificity to probe for amygdalin on a cell-by-cell basis. Methods which directly locate the unhydrolysed glycoside are clearly preferable to methods based upon detecting its breakdown products (e.g. HCN, benzaldehyde). The latter are considered more prone to problems of diffusion and thus decreased resolution. The possibility of localizing amygdalin by FT-IR microspectroscopy is currently being evaluated.

Investigation of Spatial and Temporal Regulation of Cyanogenesis in *Prunus serotina* using Molecular Approaches - An Outlook

Enzymological and immunocytochemical studies have clearly established the intriguing tissue distribution of AH, PH and MDL within mature black cherry fruits (*29, 39*). With N-terminal sequence data and specific polyclonal antibodies already in hand, we are in excellent position to investigate the developmental and tissue-specific expression of AH, PH and MDL in *Prunus* at the molecular level. To this end, we have utilized poly(A)$^+$ RNA, isolated from mid-maturation cherry seeds (35-49 DAF), to construct a cDNA expression library in λgt11 by standard procedures. This developmental stage was chosen as mRNA source because levels of all three catabolic enzymes rise dramatically 42 DAF (Figures 5 & 6). This library is currently being screened for suitable AH, PH and MDL cDNA clones using monospecific polyclonal antisera. We are optimistic that access to such clones will: (1) allow comparison of their nucleotide sequences with those of functionally related proteins (e.g. white clover linamarase (*42*), (2) further elucidate the microheterogeneity shown by AH, PH and MDL, and (3) provide molecular probes to begin investigating AH, PH and MDL gene expression in maturing black cherry fruits and in leaves. In particular, we are eager to learn the nature of *cis*- and *trans*-acting elements mediating the organ-, tissue- and cell-specific expression of these genes.

Preliminary data appear promising. Screening our library with anti-MDL antiserum has yielded 12 positive clones per 10^5 plaque-forming units. After three successive rounds of antibody screening, we now possess 2 putative MDL cDNA clones with inserts of 1.7 and 1.9 kb (cf. 1.4 kb expected for the coding region of deglycosylated MDL (50.9 kD)). The 1.9 kb insert, which hybridized with labeled oligonucleotide probes based on the MDL N-terminus, was subcloned into the Bluescript plasmid for sequencing by the dideoxynucleotide chain-termination method. Over 900 nucleotides have already been sequenced using the T_3 and T_7 primers. With the latter primer, the observed nucleotide sequence included that which encodes the known MDL N-terminus, thus confirming that the larger insert represents an authentic MDL cDNA clone. Unidirectional deletion methods are being employed to generate transformants which will allow complete sequencing of both strands.

The above library has also been screened using unpurified polyclonal antibodies raised against deglycosylated AH (this antiserum cross reacts weakly with PH). Three successive rounds of screening yielded 10 positive cDNA clones. To identify the nature of these clones, they were analysed by recently described procedures (43), in which liquid cultures of E. coli Y1090R⁻ are infected with individual λgt11 clones and induced to produce fusion proteins. Western blot analysis showed that nine of the above clones produced fusion proteins that were immunoreactive with both monospecific, affinity-purified anti-AH antibodies as well as anti-β-galactosidase antibodies. Several AH cDNA clones have been selected for sequencing.

Acknowledgments

I am especially grateful to I-Ping Cheng, Gary W. Kuroki, Chun Ping Li, Elisabeth Swain, Hua Cheng Wu and Robert Yemm who obtained most of the data described here. This research was supported by National Science Foundation Grants PCM 8314330, DCB 8917176 and IBN-9218929. Special thanks are due to Lin Anthony, Mark Woerner and Shelley Plattner for excellent assistance in manuscript preparation.

Literature Cited

1.   Conn, E. E. In *The Biochemistry of Plants: a Comprehensive Treatise; Secondary Plant Products*; Stumpf, P. K.; Conn, E. E., Eds; Academic Press, New York, 1981; Vol 7, pp 479-500.
2.   Poulton, J. E. *Plant Physiol.* **1990**, *94*, 401-405
3.   Poulton, J. E. In *Food Proteins*, Kinsella, J. E.; Soucie, W.G., Eds; The American Oil Chemists' Society, Champaign, 1989, pp 381-401.
4.   Liebig, J.; Wöhler, F. *Annalen* **1837**, *22*, 11-14
5.   Armstrong, H. E.; Armstrong, E. F.; Horton, E. *Proc. Roy. Soc. B.* **1912**, *85*, 359-362
6.   Haisman, D. R.; Knight, D. J.; *Biochem. J.* **1966**, *103*, 528-534
7.   Haisman, D. R.; Knight, D. J.; Ellis, M. J. *Phytochem.* **1967**, *6*, 1501-1505
8.   Lalégerie, P. *Biochemie* **1974**, *56*, 1163-1172
9.   Helferich, F.; Kleinschmidt, T. *Hoppe-Seyler's Z. Physiol. Chem.* **1965**, *340*, 31-36

10. Grover, A. K.; MacMurchie, D. D.; Cushley, R. J. *Biochim. Biophys. Acta* **1977**, *482*, 98-108
11. Nahrstedt, A. Dissertation, Freiburg University, FRG, 1971.
12. Nahrstedt, A. *Phytochem.* **1973**, *12*, 1539-1542
13. Kuroki, G. W.; Lizotte, P. A.; Poulton, J. E. *Z. Naturforsch.* **1984**, *39c*, 232-239
14. Kuroki, G. W.; Poulton, J. E. *Arch. Biochem. Biophys.* **1986**, *247*, 433-439
15. Li, C. P.; Swain, E.; Poulton, J.E. *Plant Physiol.* **1992**, *100*, 282-290
16. Edge, A S.; Faltynek, C. R.; Hof, L.; Keichert, L. E. Jr.; Weber, P. *Anal. Biochem.* **1981**, *118*, 131-137
17. Kuroki, G. W.; Poulton, J. E. *Arch. Biochem. Biophys.* **1987**, *255*, 19-26
18. Yemm, R. S.; Poulton, J. E. *Arch. Biochem. Biophys.* **1986**, *247*, 440-445
19. Wu, H. C.; Poulton, J. E. *Plant Physiol.* **1991**, *96*, 1329-1337
20. Shibata, Y.; Nisizawa, K. *Arch. Biochem. Biophys.* **1965**, *109*, 516-521
21. Hoesel, W.; Nahrstedt, A. *Hoppe-Seyler's Z. Physiol. Chem.* **1975**, *356*, 1265-1275
22. Hoesel, W.; Tober, I.; Eklund, S. H.; Conn, E.E. *Arch. Biochem. Biophys.* **1987**, *252*, 152-162
23. Selmar, D.; Lieberei, R.; Biehl, B.; Voigt, J. *Plant Physiol.* **1987**, *83*, 557-563
24. Goodman, M. M.; Stuber, C. W. In *Isozymes in Plant Genetics and Breeding*; Tanksley, S. D.; Nevis, D. J., Eds; Elsevier Science Publishing Co., New York, 1983; Part B, pp 1-33.
25. Brink, K. R. MS Thesis, University of California, Davis, USA, 1990.
26. Dement, W. A.; Mooney, H. A. *Oecologia* **1974**, *15*, 65-76
27. Frehner, M.; Scalet, M.; Conn, E. E. *Plant Physiol.* **1990**, *94*, 28-34
28. Majak, W.; McDiarmid, R. E.; Hall, J. W. *J. Anim. Sci.* **1981**, *61*, 681-686
29. Swain, E.; Li, C. P.; Poulton, J. E. *Plant Physiol.* **1992**, *98*, 1423-1428
30. Poulton, J. E. In *Cyanide Compounds in Biology: Ciba Foundation Symposium No. 140*; Evered, D.; Harnett, S., Eds; John Wiley & Sons, Chichester, UK, 1988, pp 67-91.
31. Müller, E.; Nahrstedt, A. Society of Chemical Ecology Meeting Abstract, Dijon, July 1991
32. Kojima, M.; Poulton, J. E.; Thayer, S. S.; Conn, E. E. *Plant Physiol.* **1979**, *63*, 1022-1028
33. Saunders, J. A.; Conn, E. E. *Plant Physiol.* **1978**, *61*, 154-157
34. Thayer, S. S.; Conn, E. E. *Plant Physiol.* **1981**, *67*, 617-622
35. Frehner, M.; Conn, E. E. *Plant Physiol.* **1987**, *84*, 1296-1300
36. Kojima, M.; Iwatsuki, N.; Data, E. S.; Villegas, C. D. V.; Uritani, I. *Plant Physiol.* **1983**, *72*, 186-189
37. Mkpong, O. E.; Yan, H.; Chism, G.; Sayre, R. T. *Plant Physiol.* **1990**, *93* 176-181
38. Selmar, D.; Frehner, M.; Conn, E. E. *J. Plant Physiol.* **1989**, *135*, 105-109
39. Swain, E.; Li, C. P.; Poulton, J. E. *Plant Physiol.* **1992**, *100*, 291-300
40. Brunswick, H. *Akademie Wissenschaften Wien* **1921**, *130*, 383-436

41.   Saunders, J. A.; Conn, E. E.; Lin, C. H.; Stocking, C. R. *Plant Physiol.*
      1977, *59*, 647-652
42.   Oxtoby, E.; Dunn, M. A.; Pancoro, A.; Hughes, M. A. *Plant Mol. Biol.*
      1991, *17*, 209-219
43.   Runge, S.W. *Biotechniques* **1992**, *12*, 630-831

RECEIVED January 12, 1993

Chapter 13

Apoplastic Occurrence of Cyanogenic β-Glucosidases and Consequences for the Metabolism of Cyanogenic Glucosides

Dirk Selmar

Botanical Institute, Technical University, Mendelssohnstrasse 4, PF 3329, D–3300 Braunschweig, Germany

Like other secondary plant products, cyanogenic glucosides are metabolized and transported in the intact plant. ß-glucosidases, capable of hydrolyzing cyanogenic glucosides at high rates occur in all parts of the plants. Generally, these ß-glucosidases are - at least partially - localized in the apoplast. Whereas the apoplastic localization of ß-glucosidases obviously separates these hydrolytic enzymes from their substrates, it excludes any occurrence of cyanogenic (mono-)glucosides in the apoplast during translocation. In contrast, the apoplastic enzymes do not hydrolyze the corresponding diglucosides, which often can be detected in addition to the cyanogenic monoglucosides, e.g. linamarin co-occurs with linustatin and prunasin with amygdalin. It is postulated that cyanogenic diglucosides act as transport metabolites, because they are protected against hydrolysis by apoplastic ß-glucosidases.

Many natural products which accumulate in plants consist of sugar moieties attached by ß-glycosidic linkage to certain aglycones (e.g. coumarinyl glycosides, glucosinolates, cyanogenic glycosides). In the same plant, ß-glucosidases capable of hydrolyzing these natural products are present. Whereas the ß-glucosides may not exhibit biological activity, the corresponding aglycones - due to their toxicicity - frequently are important ecological factors (1). Thus, the ß-glucosidases also can play an important ecological role. But, in order to evaluate the general significance of these enzymes, especially their effects on plant metabolism, in addition to biochemical data, various investigations are needed which focus on plant physiological aspects. Thus, it is necessary, to investigate both, activity and properties of the enzymes and the composition and quantities of their natural substrates in various stages of plant development.

0097–6156/93/0533–0191$06.00/0

The Significance of Compartmentation

In intact plants, ß-glucosides are usually not hydrolyzed spontaneously, due to the spatial separation of glucosidases and substrates, e.g. by the accumulation of ß-glucosides in the main vacuole which is free of the corresponding ß-glucosidases. These hydrolytic enzymes can be located in the cytoplasm or in the apoplast. The compartmentation of hydrolytic enzymes and glycosidic compounds is well documented in several examples (2-4). In the course of decompartmentation, e.g. due to tissue injury, enzymes and substrates come into contact, and the ß-glucosides are hydrolyzed. In this action, the harmless ß-glycosides can be converted directly into biologically active aglycones, or they react further to yield toxic substances. In this manner, due to the action of ß-glucosidases, glucosinolates liberate mustard oils, coumarinyl glucosides are converted to coumarins, or prussic acid is produced from cyanogenic glucosides in injured plants (Figure 1).

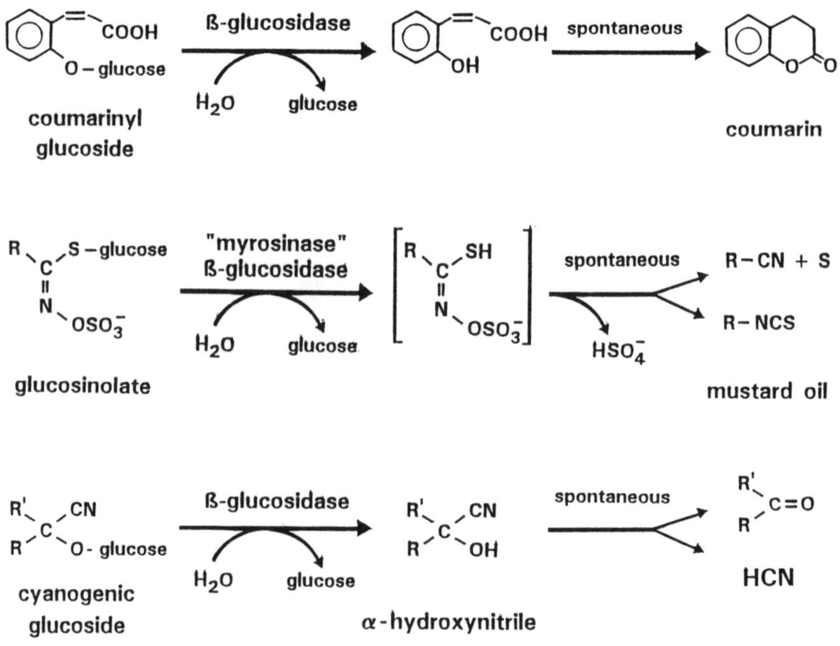

Figure 1: Cleavage of natural glucosidic products by ß-glucosidases. The unstable aglycones produced react spontaneously to form toxic compounds.

In contast to these processes involving injury, many plant glucosides are also metabolized *in vivo*. In order to prevent any intoxication of the living cell, uncontrolled cleavage of the ß-glycosidic linkage has to be avoided during the metabolism of these compounds. In this paper, the complex processes of transport and utilization of cyanogenic glucosides are presented as a model system to describe the general metabolic consequences of the apoplastic localization of ß-glucosidases.

The Metabolism of Cyanogenic Glucosides

The Metabolism of Linamarin in *Hevea brasiliensis*. In this section, fundamental aspects of the metabolism of cyanogenic glucosides are outlined by presenting some data on the transport of linamarin and its conversion to non-cyanogenic compounds. This process was studied intensively in seedlings of *Hevea brasiliensis* and some data are also available from experiments with flax plants (*Linum usitatissimum*).

Changes of HCN-Potential during Seedling Development. In the seeds of *Hevea*, large quantities of the cyanogenic glucoside linamarin are stored in the endosperm. In the course of seedling development, the total linamarin content (= HCN-potential) decreases greatly (5, 6). But, during this process, no HCN is liberated from the seedlings, meaning that the decrease in cyanogenic glucosides is not due to cyanogenesis, but instead results in the conversion of cyanogenic gluco-sides to non-cyanogenic compounds. In the course of this process the cyano-group has to be modified. Two enzymes are known to detoxify HCN in plants: rhodanese and ß-cyanoalanine synthase (7). Whereas only negligible activities of rhodanese occur in seedlings of *Hevea* (8), ß-cyanoalanine synthase is present in high amounts in the young growing leaves and roots of the seedling (6). Therefore, in order to be metabolized, the cyanogenic glucosides stored in the endosperm have to be exported into the leaves and roots of the seedling. As endosperm and cotyledons belong to genetically different organs, neither do connecting plasmodesmata exist between these two tissues nor does any vesicular transport occur. Thus, all substances translocated from the endosperm into the cotyledons have to pass the apoplastic space.

The Linustatin-Pathway. In *Hevea*, high activities of ß-glucosidase are located in the apoplastic space (9). As this enzyme readily hydrolyzes linamarin, any occurrence of intact linamarin in the apoplast must be excluded. This can be demonstrated easily by the massive liberation of HCN, when exogenous linamarin is applied onto intact endosperm or cotyledons of *Hevea* seeds (10). Kinetic analysis revealed that the apoplastic ß-glucosidase exhibits no pronounced substrate specificity (11), and a wide range of monoglucosidic substrates were hydrolyzed by this enzyme. However, the apoplastic ß-glucosidase does not act on diglucosides. These data suggested that the diglucoside linustatin is transported via the apoplast instead of the cyanogenic monoglucoside linamarin. Indeed, in contrast to the cyanogenesis resulting from linamarin application, no HCN is liberated when linustatin is applied onto intact endosperm or cotyledons of *Hevea* seeds (10). Moreover, it was found that linustatin is exuded from the endosperm of *Hevea* seeds exactly at that developmental period when linamarin is metabolized (12). Based on these data, the "linustatin-pathway" was postulated (6). According to this hypothesis (Figure 2), linamarin stored in the endosperm is glucosylated to yield the diglucoside linustatin, which then is transported via apoplastic space into the cotyledons and further is imported into the primary leaves, where it is split by the action of diglucosidase, acting by a simultaneous mechanism (13). The HCN corresponding to the acetone cyanohydrin formed, is refixed by ß-cyanoalanine synthase (6). Then, the ß-cyanoalanine produced can be hydrolyzed to asparagine, catalyzed by asparagine hydrolase (14).

Translocation of Linustatin. Although the HCN-potential decreases during the development of seedlings of *Hevea brasiliensis* "wild-type", it does not change significantly in seedlings of various other varieties of *Hevea brasiliensis* and of other *Hevea* species (15). In these cases, the same amount of cyanogenic glucosides is present in the growing leaves at the end of development as formerly accumulated

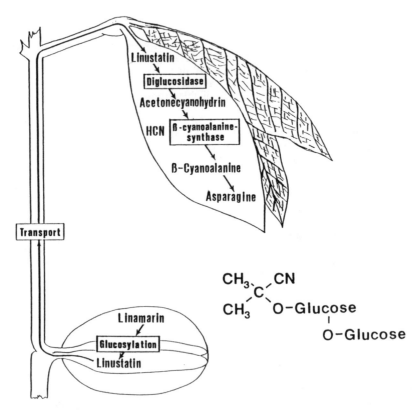

Figure 2: The Linustatin-Pathway. Linamarin, which stored in the endosperm of *Hevea* seeds is mobilized by glucosylation to yield linustatin. The cyanogenic diglucoside is translocated into the growing leaves and metabolized. Reproduced with permission from (6).

in the endosperm. This indicates that cyanogenic glucosides are simply transferred from the storage organ (source) into the young leaves (sink). As mentioned before, linamarin cannot occur in the apoplastic space. Thus, it is very likely that linustatin is not only involved in the mobilization and utilization of linamarin but also in the simple translocation of linamarin.

In order to monitor the metabolic fate of linustatin, as it is transferred from the endosperm into the cotyledons, ^{14}C-linustatin was applied to cotyledons of *Hevea*. ^{14}C-linustatin was taken up very efficiently, supporting the hypothesis that linustatin indeed acts as transport metabolite (10). When ^{14}C-linustatin is taken up initially, most of the radiolabel is still present within the imported diglucoside. Some hours later, large amounts of the radioactivity are present as ^{14}C-linamarin, and additionally ^{14}C was also incorporated into non-cyanogenic compounds. During futher incubation, ^{14}C-linamarin as well as ^{14}C-labelled non-cyanogenic compounds are present in the growing leaves. These data verify that linustatin is involved in both processes, the translocation of linamarin and the conversion to non-cyanogenic compounds (10). The complex interaction of linamarin and linustatin and the possible conversions are indicated in Fiure 3.

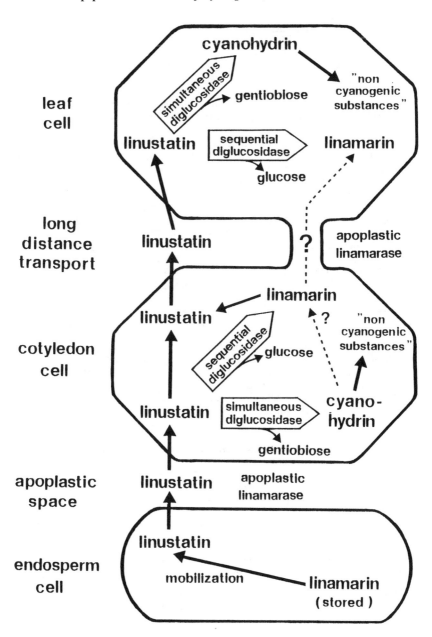

Figure 3: Interaction of interconversion of cyanogenic mono- and diglucosides. Linustatin exported out of the cotyledons either can be converted back to linamarin or can be metabolized to non-cyanogenic compounds.

Lotaustralin and Neolinustatin - an Analogous System. In some varieties of *Hevea brasiliensis* as well as in some other *Hevea* species, another cyanogenic monoglucoside, lotaustralin, occurs in small amounts (*16*) together with large amounts of linamarin. Since the apoplastic linamarase also hydrolyzes lotaustralin (*12*), the conclusions deduced for linamarin also should be valid for lotaustralin. Thus, the diglucoside corresponding to lotaustralin, neolinustatin, should also act as a transport metabolite. Indeed, in all lotaustralin-containing seedlings of *Hevea* spp., neolinustatin is detectable parallel to the occurrence of linustatin (*15*).

Apoplastic ß-Glucosidase in Various Cyanogenic Plants. The question arises whether the interaction of cyanogenic mono- and diglucosides is unique for *Hevea* or whether diglucosidic cyanogens act as transport metabolites for the corresponding monoglucosides in other plants.

In order to study this problem, one has to ask if the ß-glucosidases which hydrolyze cyanogenic glucosides in other cyanogenic plants occur apoplastically. The apoplastic occurrence of such ß-glucosidases would appear to exclude the presence of cyanogenic monoglucosides in the apoplastic space.

The occurrence of ß-glucosidases in the apoplastic space in general is well established (*3, 17, 18*) and a lot of quite different apoplastic ß-glucosidases have been investigated (i.e *19, 20*). More recently, the work of Marcinowski & Grisebach (*21, 22*) showed that the glucosidic lignin precursors (e.g. coniferin and syringin) are hydrolyzed by ß-glucosidases localized within the cell wall during lignification. It is well accepted that lignification processes are very similar in all higher plants, and therefore, at least those ß-glucosidases are present extraplasmatically, which are able to hydrolyze coniferin and syringin.

Several studies concerning apoplastic localization of cyanogenic ß-glucosidases, specifically on linamarin-hydrolyzing enzymes (linamarases), are available in the literature (Table I). In all plants tested, high activities of linamarase was present in the apoplastic space. The apoplastic localization of linamarase was documented by histological or immunocytochemical means as well as by analyses of extracellular liquids and isolated protoplasts (Table I).

Table I: Apoplastic Occurrence of ß-Glucosidases

Plant Species	Method	References
Lotus corniculatus	histochemistry	Rissler & Millar (23)
Trifolium repens	intercellular liquid	Kakes (24)
Phaseolus lunatus	isolated protoplasts	Frehner & Conn (25)
Manihot esculenta	intercellular liquid	Kurzhals et al. (27)
Dimorphotheca sinuata	intercellular liquid	Grützmacher (26)
Hevea brasiliensis	isolated protoplasts	Kurzhalz et al. (27)
Manihot esculenta	immunochemistry	Mkpong et al. (28)

In general, the presence of an apoplastic linamarase in leaves can easily be demonstrated by infiltration of linamarin (*25*): when linamarin enters the extracellular spaces of the leaves, it is hydrolyzed and HCN is liberated, produced by the dissociation of the aglycone (acetone cyanohydrin). The absence of cyanogenesis in controls, where only water is infiltrated, proves that the liberation of HCN indeed is due to the cleavage of exogenous linamarin. In contrast to this, no HCN is liberated when the cyanogenic diglucoside linustatin is infiltrated into leaves of *Manihot esculenta, Hevea brasiliensis* or *Dimorphotheca sinuata*. This means that linustatin cannot be hydrolyzed by apoplastic enzymes in these species. In order to test this hypothesis, we analyzed extracellular liquids, which were obtained according to Frehner & Conn (*25*), with special regards to the presence of linamarase and linustatinase (Table II).

Table II. Apoplastic Linamarase Activities

Plant Species	Enzyme Activity [a] (nmol x min^{-1} x g FW^{-1}) (%)					
	Linamarase		Linustatinase		Malate dehydrogen.	
	homogen.	apoplast.	homogen.	apoplast.	homogen.	apoplast.
Dimorphotheca sinuata	699	8 (0,4-1,1)	n.d.	0,0 (0,0)	8	0,0 (0,0)
Hevea brasiliensis	5200	1760 (21-53)	n.d.	0,0 (0,0)	9490	0,0 (0,0)
Linum usitatissimum	790	409 (30-42)	n.d.	0,0 (0,0)	6430	0,0 (0,0)
Lotus corniculatus	413	2 (0,5-49)	n.d.	0,0 (0,0)	9	0,0 (0,0)
Manihot esculenta	3380	2100 (52-73)	n.d.	0,0 (0,0)	15	0,0 (0,0)
Phaseolus lunatus	912	618 (35-45)	n.d.	0,0 (0,0)	9915	0,0 (0,0)
Trifolium repens	162	3 (1,8-2,9)	n.d.	0,0 (0,0)	4	0,0 (0,0)

a) The absolute enzyme activities correspond to one experiment. In total, two to six of such experiments were performed. The relative values, which mention the recoveries of the enzyme activities in the apoplastic fluids (apoplast.) in relation to the corresponding hypotonic leaf homogenates (homogen.), show the variations within the different experiments. Linamarase and linustatinase were determined according to (*11*). Malate dehydrogenese (malate dehydrogen.), used as marker for cytosolic enzymes, was measured as described in (*25*). Due to the presence of large amounts of linamarin, linustatinase could not be determined (n.d.) in the homogenates. Apoplastic fluids were obtained according to (*25*).

Linamarase activity was detectable in the extracellular liquids of all plants tested, (Table II), supporting the literature data on the occurrence of apoplastic linamarase (Table I). In contrast, none of these samples contained enzymes, which were capable of hydrolyzing linustatin. Thus, in intact tissues of these plants the apoplastic occurrence of linamarin must be excluded, whereas linustatin indeed can be present in the apoplast.

The occurrence of linustatin in linamarin containing plants. If linustatin functions generally as transport metabolite, this cyanogenic diglucoside should be present also in other linamarin containing plants. Linustatin was first described and isolated from flax seeds (29). The authors also detected the lotaustralin-glucoside, neolinustatin. The first report of linustatin in vegetative tissues was by Spencer et al. (30), who detected linustatin in leaves of *Passiflora pendens*. Subsequently, the occurrence of linustatin has been demonstrated for several linamarin containing plants, and indeed, in all plants so far analyzed, linustatin, at least in traces, is detectable (Table III).

Table III: Occurrence of Linustatin in Linamarin-containing Plants

Plant Species	References
Linum usitatissimum	Smith et al. (29)
Passiflora pendens	Spencer et al. (30)
Hevea brasiliensis	Selmar et al. (12)
Phaseolus lunatus	Frehner et al. (31)
Dimorphotheca sinuata	Frehner et al. (31)
Acacia farnesiana	Frehner et al. (31)
Trifolium repens	Kurzhals et al. (9)
Manihot esculenta	Kurzhals et al. (9)

Translocation of Cyanogenic Glucosides.

In the previous sections, the general apoplastic localization of linamarin was excluded and, in addition, the corresponding diglucoside linustatin shown to be present in all linamarin containing plants. Thus, it is proposed that in these plants, analogous to *Hevea*, translocation of cyanogenic glucosides also involves the cyanogenic diglucoside as the transport metabolite.

Evidence for the Transport of Cyanogenic Glucosides. Translocation of cyanogens has been investigated in only a few cases. Clegg et al. (32) concluded from reciprocal changes in HCN-potential (decrease in the cotyledons and an increase in the true leaves) that in *Phaseolus lunatus* linamarin is transported from the storage cotyledons into the growing leaves. Quite similar changes, detectable in seedlings of *Dimorphotheca sinuata*, imply that cyanogens also are translocated from one leaf generation to another in this plant (33). De Bruijn (34) postulated a transport of cyanogenic glucosides in *Manihot esculenta* based on ringing experiments with stems of cassava, resulting in an increase of the HCN-potential above the incision. This conclusion was confirmed by analyzing the distribution of ^{14}C-labelled linamarin in cassava plants after precursor feeding (*35*) and by reciprocal stem grafts (*36*).

In the case of *Hevea* seedlings, the apoplastic transport of cyanogens is required because all substances transported from endosperm to the cotyledons have to pass the apoplastic space between these two organs. When cyanogens are translocated from old leaves (source) into younger ones (sink), there is of course no endosperm-cotyledon passage involved. But during long distance transport, apoplastic steps should be involved: in the course of phloem transport, this is due to the apoplastic phloem loading, e.g. (*37*), whereas xylem transport is entirely apoplastic. Thus, when cyanogens are transported via the long distance transport systems, the substances would pass apoplastic space. As discussed above, cyanogenic mono-glucosides cannot occur in the apoplastic space due to the presence of ß-glucosi-dases. Thus, when cyanogenic monoglucosides are translocated within the plant via the vascular system, the translocation involves the transport of the corresponding diglucosides.

Translocation of Cyanogens from Endosperm into Cotyledons. The translocation of cyanogenic glucosides within adult plants is mainly deduced from the overall changes in HCN-potential, but no data on the qualitative composition of cyanogenic glucosides are available. However, changes in the ratio of mono- and diglucosides which occur in the course of the maturation of several cyanogenic seeds are well documented. In immature seeds of several species of the Rosaceae, the HCN-potential is due to the monoglucoside prunasin, whereas in mature seeds the corresponding diglucoside is exclusively accumulated (*Cotoneaster bullata:* (*38*), *Prunus amygdalus*: (*39*). Frehner et al. (*39*) found similar changes in the pattern of cyanogenic glucosides in developing flax seeds: in immature seeds, they detected the monoglucosides linamarin and lotaustralin exclusively whereas in mature seeds only the corresponding diglucosides linustatin and neolinuststin were present. The authors explain these changes in ratio by glucosylation of mono-glucosides to yield the corresponding diglucosides, but these data have not been evaluated in regard to translocation processes.

Production and translocation of Linustatin in Flax Seeds. To understand these processes it is necessary to consider the morphology of developing flax seeds and combine them with the biochemical data. When flax seeds are fully grown but still immature, endosperm comprises the major part of the seed and the embryo is still very small (Figure 4). At this developmental stage, the seeds contain only linamarin and lotaustralin (*38*), thus the cyanogenic monoglucosides are localized in the endosperm. After one week, the seeds contain no cyanogenic mono-glucosides, and only the corresponding diglucosides are present, accumulated in the big storage cotyledons. Thus, conversion of monoglucosides to the corresponding diglucosides is accompanied by the translocation of cyanogens from endosperm to cotyledons. The synthesis of linustatin from linamarin has been proved by feeding experiments with ^{14}C-labelled precursors (10, Selmar, *1993, Planta in press*). If we consider that apoplastic linamarase excludes the occurrence of linamarin in the apoplastic space between the endosperm and cotyledons, these data prove that also in *Linum usitatissimum* - analogous to the *Hevea* system - linamarin is translocated via linustatin as the transport metabolite, protected against cleavage by apoplastic enzymes.

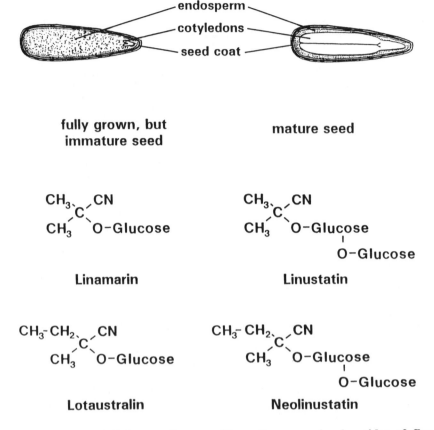

Figure 4. Morphology and composition of cyanogenic glucosides of flax seeds (*Linum usitatissimum*).

Apoplastic ß-glucosidase *in Prunus* ssp. For seeds of several *Prunus* species conditions seem different from the *Hevea* system, because the prunasin hydrolyase is localized exclusively within the plant cell (Poulton, previous chapter in this book). However, we were able to detect apoplastic ß-glucosidase also in several *Prunus* seeds, which was also capable of hydrolyzing prunasin efficiently. This is demonstrated by direct incubation of endosperm with cyanogenic substrates (Figure 5). When prunasin is applied to the endosperm tissue (uninjured), the cyanogenic glucoside is hydrolyzed quickly, detected by the release of HCN. In contrast, amygdalin is not cleaved by the apoplastic enzymes present in *Prunus* seeds.

The apoplastic localization of prunasin-hydrolase activity contradicts the immuno-histochemical data on prunasin hydrolyase (Poulton, previous chapter in this book), but there are two possible explanations: a) the apoplastic prunasin hydrolyase corresponds totally to the cytosolic prunasin hydrolyase, but due to its association to the cell wall, the immunolabelling is suppressed or b) in *Prunus* seeds besides the specific prunasin hydrolyase, purified by (*40*) and used to rise the antibodies, additionally an apoplastic ß-glucosidase is present, which also is able to

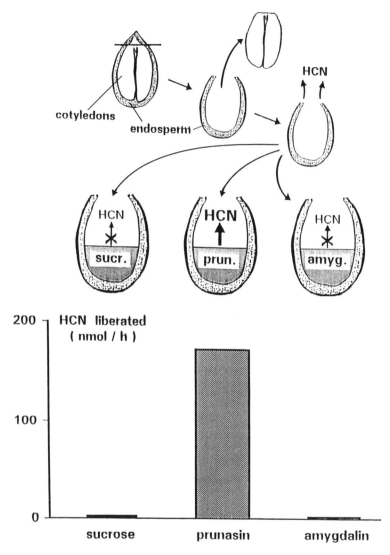

Figure 5. Apoplastic prunasin hydrolyase in seeds of *Prunus domestica*.
After removal of the cotyledons, plum seeds were incubated in Erlenmeyer
flasks. Since the cotyledons were removed very carefully, only those injuries
result, which are due to the incision. In contrast, the major part of the
endosperm is intact and not injured. Some hours later, when wound induced
cyanogenesis stopped, 20 μl of 10 mmol/l prunasin or 10 mmol/l amygdalin
prunasin were applied to the endosperm. In order to prevent damage due to
osmotic stress, solutions contained also 200 mmol/l sucrose. Sucrose solution
(without cyanogens) was applied as control. Due to the action of apoplastic
enzymes, the cyanogenic glucosides would be split. To trap the HCN released,
air was streamed through the flasks and bubbled through NaOH solution.

hydrolyze prunasin efficiently. It is well known that, in addition to specific ß-glucosidases, more *general ß-glucosidases* do occur in plants, which reveal only little substrate specificity. Thus, in many cyanogenic plants, several ß-glucosidases occur which differ significantly in their catalytic properties (e.g. *41-43*). In apple seeds, belonging like *Prunus* to the Rosaceae family, Podstolski & Lewak (*44*) found at least three different ß-glucosidases. It is therefore very likely that in *Prunus* seeds, an additional ß-glucosidase occurs apoplastically, that presumably is necessary for the cleavage of lignin precursors.

   Production and Translocation of Amygdalin in *Prunus* ssp. To understand the interaction of prunasin and amygdalin, one again has to consider the morphology of developing *Prunus* seeds and combine this information with the biochemical data. A young *Prunus* seed is nearly totaly filled with a large body of endosperm, while the embryo is very small (Figure 6). At this developmental stage, the seeds contain only prunasin (*39*), which is localized in the endosperm. In mature seeds, where no cyanogenic monoglucosides are detectable, amygdalin is accumulated in the large storage cotyledons. Thus, in a manner analogous to flax seeds, the conversion of monoglucosides to the corresponding diglucosides is accompanied by the translocation of cyanogens from endosperm to cotyledons. If we consider that the apoplastic ß-glucosidase excludes the occurrence of prunasin

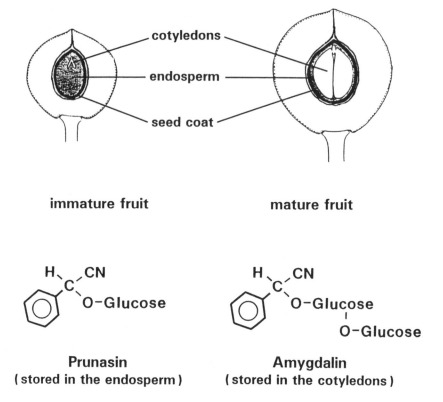

Figure 6: Morphology and composition of cyanogenic glucosides of cherry seeds (*Prunus* ssp).

in the apoplastic space between endosperm and cotyledons, it has to be concluded that in *Prunus* - analogous to the *Hevea* system - prunasin is not transported directly but is translocated as the corresponding diglucoside: amygdalin acts as transport metabolite, protected against hydrolysis by the apoplastic enzymes. These examples indicate that translocation analogous to the "linustatin-pathway" is not restricted to few special plants, but seems to be a more general feature of cyanogenic plants.

Apoplastic ß-Glucosidases - General Aspects.

Lignification is a general feature of higher plants. Apoplastic ß-glucosidases, presumably involved in this process, occur in all higher plants. In many plants, in addition to these enzymes, other ß-glucosidases, which are responsible for the *post mortem* hydrolysis of several glucosidic natural products (e.g. coumarinyl glycosides, glucosinolates, cyanogenic glycosides) are present. These hydrolytic enzymes, which very often exhibit high substrate specificity (e.g. Conn, in this volume), are localized within the plant cell (*19, 43*) or in the apoplast (*2, 24, 25, 44*). In contrast, in plants like *Hevea brasiliensis*, where only one ß-glucosidase is detectable, this enzyme is exclusively apoplastic and does not exhibit pronounced substrate specificity.

Despite the varying properties, in all cyanogenic plants tested, apoplastic ß-glucosidases are able to cleave endogenous cyanogenic monoglucosides, whereas the corresponding diglucosides are resistant to hydrolysis. This means, diglucosidic substrates, in contrast to monoglucosidic ones, can pass the apoplast without being hydrolyzed. It is postulated that generally, when cyanogenic glucosides are translocated within the plant, the corresponding cyanogenic diglucosides function as transport metabolites. It has to be investigated to see whether this process applies to other natural glucosidic subtrates.

In addition to the general assumption that glucosylation of several compounds is required in order to enhance their solubility, we postulate that a second glucosylation step (to yield diglucosidic compounds) is required for translocation of glucosides in order to pass though the apoplstic space, e.g. required in transport from endosperm to the cotyledons or in the course of phloem loading.

Literature cited

(1) Nahrstedt, A. *Plant Syst. Evol.* **1985**, *105*, 35-47
(2) Kojima, M., Poulton, J. E., Thayer, S. S., Conn, E. E. *Plant Physiol.* **1979**, *67*, 617-622
(3) Matile, Ph. *The Lytic Compartment of Plant Cell,* Cell Biology Monographs **1975**, Springer Verlag
(4) Oba, K., Conn, E. E., Canut, H., Boudet, A. M. *Plant Physiol.* **1981**, *68*, 1359-1363
(5) Lieberei, R., Selmar, D., Biehl, B. *Plant Syst. Evol.* **1985**, *105*, 49-63
(6) Selmar, D., Lieberei, R., Biehl, B. *Plant Physiol.* **1988**, *86*, 711-716
(7) Miller, J. M., Conn, E. E. *Plant Physiol.* **1980**, *65*, 1199-1202
(8) Lieberei, R., Selmar, D. *Phytochem.* **1990**, *29*, 1431-1434
(9) Kurzhals, Ch., Grützmacher, H., Selmar, D., Biehl, B. *Planta Medica* **1989**, *55*, 673
(10) Selmar, D. 1992. *Habilitationsschrift,* **1992,** Naturwissenschaftliche Fakultät Technische Universität, Braunschweig
(11) Selmar, D., Lieberei, R., Biehl, B., Voigt, J. *Plant Physiol.* **1987**, *83*, 557-563

(12) Selmar, D., Lieberei, R., Biehl, B. Nahrstedt, A., Schmidtmann, V., Wray, V. *Phytochem.* **1987**, *30*, 2135-2140
(13) Kuroki, G, Lizotte, P. A., Poulton, J. E. *Plant Physiol.* **1979**, *63*, 1022-1028
(14) Castric, P. A., Farnden, K. F., Conn, E. E. *Arch. Biochem. Biophys.* **1972**, *152*, 62-69
(15) Selmar, D., Lieberei, R., Junqueira, N., Biehl, B. *Phytochem.* **1987**, *26*, 2400-2401
(16) Lieberei, R., Nahrstedt, A., Selmar, D., Gasparotto, L. *Phytochem.* **1986**, *25*, 1573-1578
(17) Rothstein, S. J. *The Enzymology of the Cell Surface*; Protoplasmalogia II, E4, Springer
(18) Heyn, A. N. J. *Arch. Biochem. Biophy.* **1969**, *132*, 442-449
(19) Keegstra, K., Albersheim, P. *Plant Physiol.* **1970**, *45*, 675-678
(20) Matile, Ph. *Biochem. Physiol. Pflanzen* **1980**, *175*, 722-731
(21) Marcinowski, S., Grisebach, H. *Eur J. Biochem.* **1978**, *87*, 37-44
(22) Marcinowski, S., Falk, H., Hammer, D. K., Hoyer, B., Grisebach, H. *Planta* **1979**, *144*, 161-167
(23) Rissler, J. F., Millar, R.L. Protoplasma **1977**, *92*, 57-70 Verlag, Wien
(24) Kakes, P. *Planta* **1985**, *166*, 156-160
(25) Frehner, M., Conn, E. E. *Plant Physiol.* **1987**, *84*, 1296-1300
(26) Grützmacher, H., *Diploma thesis*, **1989**, Naturwissenschaftliche Fakultät Technische Universität, Braunschweig
(27) Kurzhals, Ch., Selmar, D., Biehl, B. *Tagung der Deutschen Botanischen Gesellschaft, Regensburg* **1990**, P5-8
(28) Mkpong, O. E., Yan, H., Chism, G., Sayre, R. T. *Plant Physiol.* **1990**, *93*, 176-181
(29) Smith, C. R. jr., Weisleder, D., Miller, R. W. J. *Org. Chem.* **1980**, *45*, 507-510
(30) Spencer, K. C., Seigler, D. S., Nahrstedt, A. *Phytochem.* **1986**, *25*, 645-647
(31) Frehner, M., Selmar, D., Conn, E. E. *Annual Meeeting of the Phytochemical Society of North America* **1987**, *Paper 4*
(32) Clegg, D. O., Conn, E. E., Janzen, D. H. *Nature* **1979**, *278*, 343-344
(33) Grützmacher, H., Biehl, B., Czygan, F., Selmar, D. *Planta Medica* **1990**, *56*, 610 611
(34) De Bruijn, G. H. In *Chronic Cassava Toxicity*; Nestel, B., MacIntyre, R., Eds.; International Development Research Centre, Ottawa, Canada, 1973; 43-48
(35) Bediako, M. K. B., Tapper, B. A., Pritchard, G. G.; *Tropical Root Crops: Research strategies for the 1980s*; Terry, E. R., Oduro, K. A., Caveness, F., Eds.; International Development Research Centre, Ottawa, Canada; 1981; pp 143-148
(36) Makame, M., Akoroda, M. O., Hahn, S. K. *J. Agric. Sci. Camb.* **1987**, *109*, 605-608
(37) Delrot, S. *Plant Physiol. Biochem.* **1987**, *25*, 667-676
(38) Nahrstedt, A. *Phytochem.* **1986**, *12*, 1539-1542
(39) Frehner, M., Scalet, M., Conn, E. E. *Plant Physiol.* **1990**, *94*, 28-34
(40) Kuroki, G. W., Poulton, J. E. *Biochem. Biophy.* **1987**, *255*, 19-26
(41) Mao, C.-H., Anderson, L. *Phytochem.* **1967**, *6*, 473-483
(42) Nahrstedt. A., Hösel, W., Walther, A. *Phytochem.* **1979**, *18*, 1137-1141
(43) Thayer, S. S., Conn, E. E. *Plant Physiol.* **1981**, *67*, 617-622
(44) Boersma, P., Kakes, P., Schramm, A. W. *Acta Bot. Neerl.* **1983**, *32* 39-47

RECEIVED February 4, 1993

Chapter 14

Identification and Characterization of a Novel Phytohormone Conjugate Specific β-Glucosidase Activity from Maize

N. Campos, L. Bako, B. Brzobohaty, J. Feldwisch, R. Zettl, W. Boland, and K. Palme

Max-Planck Institut für Züchtungsforschung, Carl-von-Linné-Weg 10, D-5000 Cologne 30, Germany

Virtually all aspects of plant growth and development are influenced by structurally relatively simple substances termed phytohormones. It has been argued that the wide range of responses elicited by these substances requires a mode of action that is radically different from those of animal hormones. Current evidence indicates that enzymes that can synthesize and modify phytohormones and their antagonists, or hydrolyze phytohormone conjugates to release active hormones, play a role in initiating important regulatory pathways. They are also likely to provide invaluable tools with which to study the mechanisms underlying growth and differentiation in plants. Here we describe recent biochemical progress in the characterization of the molecular targets of phytohormones.

Vascular plants posess unique morphological and developmental features that can be viewed as specific adaptations to the autotrophic sessile habit of these organisms. Plant development has been shown to be influenced by a set of five phytohormones (auxins, cytokinins, gibberellins, abscisic acid and ethylene), each of which can elicit a remarkable variety of responses (for review see 1). Auxins (indole-3-acetic acid and its natural and synthetic derivatives) promote cell division, cell elongation, differentiation of vascular tissues and root initiation. They have been proposed to influence apical dominance and tropisms. Cytokinins are thought to promote shoot initiation, cell division and, together with auxins, regulate plant growth and differentiation.

Numerous physiological experiments have illustrated the important role of these phytohormones. Several unanswered questions, however, remain: (i) what are the molecular mechanisms underlying phytohormone action, (ii) what is the importance of phytohormone transport between cells and tissues, (iii) what

0097–6156/93/0533–0205$06.00/0

imbibition during the rapid growth phase. Protein extracts were prepared from these coleoptiles and after photoaffinity labeling with 5'-azido-[7-^{3}H]indole-3-acetic acid ([^{3}H]N$_3$IAA) a 60 kDa protein, termed p60, was identified. This protein was initially detected in the post-ribosomal supernatant, indicating that it might be present in the cytosol of the intact cells. p60 was also detected in protein extracts prepared after solubilization of microsomal fractions.

To demonstrate the specificity of photoaffinity labeling of p60, competition studies were performed using various unlabeled auxin analogues (13). Physiologically active natural and synthetic auxins (IAA, 1-naphthylacetic acid and 2,4-dichlorophenoxyacetic acid) significantly reduced the incorporation of [^{3}H]N$_3$IAA into p60. Compounds specific for the indole ring, such as L-tryptophan, or compounds specific for the aromatic ring system, or radical scavengers, such as p-aminobenzoic acid, did not compete the labeling of p60.

To study the presence of p60 in different organs of the maize seedling, we isolated crude microsomal fractions from coleoptiles (including the node and the primary leaf), the mesocotyl and roots. Photoaffinity labeling of the corresponding protein extracts showed that p60 was mainly present in the coleoptile fraction. It was less abundant in roots and virtually absent in the mesocotyl fraction.

The molecular mass of the native p60 was estimated by gel filtration chromatography. A cytosolic fraction containing p60 was separated by Sephacryl S-200 gel filtration and its molecular mass was determined as 130 kDa. This result suggests that p60 may dimerize. We analyzed microsomal p60 by two-dimensional gel electrophoresis and found that it separated into three different isoforms. This heterogeneity indicates the presence of either post-translational modifications or genetic variability. The functional relevance of this heterogeneity is not yet known. Interestingly, dimerization of another enzymatically active maize ß-glucosidase has been demonstrated previously (124) and 2-3 charge isomers of this enzyme were observed (15).

In addition, we found another p60-related protein (pm60) in plasma membranes (16; our unpublished observations). Plasma membranes were isolated by two-phase partitioning and shown to be highly enriched for plasma membrane vesicles by analysis of several marker enzymes. pm60 appeared to be tightly bound to the membrane as it was not readily solubilized by extraction with either n-butanol or acetone, or by extraction of the membranes with the detergents Triton X-100, CHAPS or digitonin (unpublished results). However, p60 was readily solubilized after treatment with Triton X-114, a detergent known to release integral membrane proteins from the lipid bilayer (17).

Another p60-related protein with a slightly smaller apparent molecular mass (p58) was found in microsomal fractions (unpublished results). p58 is a minor protein in maize coleoptiles and present at less than 5% of p60, as estimated by photoaffinity labeling. p58 was detected in crude extracts only after polyacrylamide gels containing photoaffinity labeled p60 were exposed to X-ray films for more than two weeks. p58 does not appear to be a degradation product of p60. We found that although it was extracted together with p60 it could be separated from p60 quantitatively by binding to carboxy-methyl-Sepharose.

The 60 kDa protein (p60) was purified to homogeneity by using several mechanisms control the intracellular and extracellular levels of active phytohormones and the sensitivity of their target cells, and (iv) how are the phytohormone sensitivity characteristics of critical physiological responses controlled? As plants lack the structures analogous to the specific endocrine organs of animals for spatially localized hormone production, it is reasonable to expect different principles to apply to the organization of the production, movement and sequestration of phytohormones. Recent insights into such processes demonstrate the involvement of various glycosidases that can metabolize phytohormones (*2,3*). Glycosidases probably alter the cellular concentrations of active forms of phytohormones and thus are likely to play an important role in the control of plant development.

Conjugation of phytohormones to low molecular weight compounds is probably part of the mechanism for control of hormone concentrations. The molecular details are only poorly understood (*4*). Why do plants need to amplify their spectrum of growth regulators through formation of numerous metabolites? These compounds are often much more abundant than the physiologically active phytohormones, but are they part of the plants strategy to downregulate hormone activity or even to degrade these compounds?   Or are these metabolites intermediates in hormone transport as some experiments suggest (*5,6*)? Alternatively, these widely distributed compounds could be part of the plants' strategy to cope with the continuous flux of environmental changes. Some of these compounds could provide the key to a molecular analysis of the regulation of plastic growth responses in plants.

The identification of cellular proteins responsible for phytohormone binding or metabolism will be an important step in the analysis of phytohormone conjugates. It will also allow the molecular details underlying the regulation of expression of the corresponding genes, for example in response to various environmental stimuli, to be elucidated.

Genetic or biochemical strategies may be applied to identify such proteins and their corresponding genes. Photoaffinity labeling techniques, which overcome problems associated with hormone-binding techniques, have in many cases proven to be successful for the identification of hormone-binding proteins and even hormone receptors (*7-12*). It can be envisaged that some of the plant proteins identified by photoaffinity labeling with light-sensitive synthetic phytohormone derivatives, may be enzymes able to metabolize phytohormones or their conjugates.

Methods

Methods were previously described (*12,13*).

Results and Discussion

Identification and Characterization of Different Isoforms of p60.   Maize coleoptiles grown in the dark at 28°C were harvested at 3 days after seed chromatographic separation steps including chromatography on Sepharose linked to 1-naphthylacetic acid, from which p60 was eluted with 1-naphthylacetic acid, a well- known synthetic auxin analogue. This resin was used earlier for the purification of another well-studied auxin-binding protein that is located in the ER of maize coleoptiles (*18-20*).

ß-Glucosidase Activity of p60. Microsequencing studies revealed that p60, p58 and pm60 are members of a protein family and were differentially processed (Table I). The primary amino acid sequence of p60 revealed similarities to ß-glucosidases (our unpublished observation); therefore p60 was analyzed for ß-glucosidase activity. It was found that a fraction containing p60 did indeed exhibit ß-glucosidase activity towards general ß-glucosidase substrates (e.g. p-nitrophenyl-glucopyranoside or 6-bromo-2-naphthyl-ß-D-glucopyranoside/fast blue BB for activity staining in native polyacrylamide gels). The results are presented in Figure 1. Specific staining of p60 with 6-bromo-2-naphthyl-ß-D-glucopyranoside/ fast blue BB demonstrated that p60 has ß-D-glucoside glucohydrolase activity (E.C. 3.2.1.21). To characterize the ß-glucosidase activity of purified p60 we performed enzyme assays with the synthetic substrate p-nitro-phenyl-ß-D- glucopyranoside and found maximal activity between pH 5.3 and 5.9. This range is in a good agreement with the pH optimum of 5.8 determined for a maize ß-glucosidase described earlier (*15, 21*).

Table I.   N-terminal amino acid sequences of pm60 and p60

---

pm60   S A R V G S Q N G V Q M L S P S E I P Q

p60   S Q N G V Q M L S P S E I P Q

We assayed the enzyme activity of the other proteins related to p60: p58 and pm60. In each case, the presence of the p60-related proteins correlated with a high ß-glucosidase activity. This suggests that the variants of p60 detected by photoaffinity labeling are ß-glucosidases.

To define the substrate specificity of p60, we tested different compounds that are commonly cleaved by a wide variety of glucosidases. For comparison, we also analyzed the substrate specificity of a ß-glucosidase from *Caldocellum saccharolyticum*, an enzyme involved in the breakdown of cellulose (*22*). In contrast to other ß-glucosidases (i.e., ß-glucosidase from *C. saccharolyticum* or emulsin from sweet almonds), which hydrolyze a broad range of substrates, p60 had a distinct pattern of substrate specificity (shown in Table II). These data suggested that p60 is unlikely to be involved in the breakdown of cellulose *in vivo* (*13*).

The results of photoaffinity labeling discussed above support the view that the p60 can bind auxins. In this context, it is important to determine whether the presence of auxins has any effect on the ß-glucosidase activity of p60. We found that both IAA and 1-naphthylacetic acid inhibit p60-associated ß-glucosidase

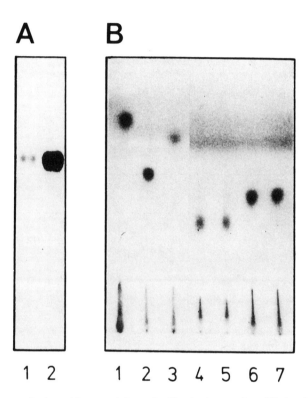

Figure 1. β-glucosidase activity of p60. **A.** 1 μg of purified p60 protein was electrophoresed on a 10% native polyacrylamide gel and either stained with Coomassie Blue (lane 1) or enzymatically stained for ß-glucosidase activity with 6-bromo-2-naphthyl-ß-D-glucoside/fast blue BB (lane 2). **B.** Indole derivatives at a final concentration of 1 mM were incubated in citrate buffer, pH 5.5 in the absence (lanes 2,4,6) or presence (lanes 3,5,7) of 0.5 U p60 protein. Products of the reaction were analyzed by TLC. Lane 1, IAA; lanes 2,3, indoxyl-ß-D-glucoside; lanes 4,5, IAA-myo-inositol; lane 6,7, IAA-L-aspartate.

activity in a competitive manner. This effect is specific since other bacterial ß-glucosidases are not affected by auxins. In addition, the presence of non-functional auxin analogues such as L-tryptophan or 5-hydroxy-IAA, or aromatic compounds, such as benzoic acid, had no effect on the ß-glucosidase activity of p60. These results suggest that IAA and related compounds are aglycones which can bind to the active site of p60. Further experiments demonstrated that p60 readily hydrolyzed indoxyl-β-glucoside, a synthetic compound structurally related to the natural auxin conjugate indole-3-acetyl-ß-D-glucoside (Figure 1). This activity appeared to be highly specific since the p60 was not able to hydrolyze other indole-acetic acid conjugates like IAA-myoinositol or IAA-L-aspartate. Present data suggest that p60 might be involved *in vivo* in the hydrolysis of glucosidic phytohormone conjugates *(13)*.

Table II.   Substrate specificities* of p60 associated ß-glucosidase activity and the ß-glucosidase from *C. saccharolyticum*

	Activity	
		(% of PNPG)
Substrate	p60	*C. saccharolyticum*
p-nitrophenyl-ß-D-glucopyranoside	100	100
p-nitrophenyl-ß-D-galactopyranoside	0	241.4
p-nitrophenyl-ß-D-arabinoside	27	9.1
phenyl-ß-D-glucopyranoside	0	324.6
methyl-ß-D-glucopyranoside	0	26.3
(ß-1,4)-cellobiose	0	245.6
(ß-1,3)-laminiaribiose	17.5	765.0

*The assay was performed for 30 min at 30°C in 100 mM citrate buffer (pH 5.5) using 5 mM of substrate. Product analysis was perfomed   spectrophotometrically as described in *Methods*. As PNPG was used by both enzymes with similar K_m values (K_m 0.4 mM for p60 and 0.6 mM for ß-glucosidase from *C. saccharolyticum*), all values obtained were normalized to the level of PNPG cleavage.

Extensive amino acid sequence analysis of this protein allowed the construction of several synthetic oligonucleotide probes which were used to isolate a cDNA clone coding for a protein related to p60. The cDNA, named *Zm-p60.1*, corresponded to a mRNA with a 3'-poly(A)$^+$ sequence and a single open reading frame. When the Zmp60.1 primary sequence was compared with other amino acid sequences available in protein data bases, similarities were observed in the range from 30 to 50% to other ß-glycosidases from archaebacteria, eubacteria and eukaryotes. Much to our surprise we found a stretch of amino acids similar to part of the rolC protein from *Agrobacterium rhizogenes*. The motif shared between the rolC protein and Zmp60.1 pointed to the possibility that both proteins could share common substrates. It is expected that transgenic tobacco plants stably expressing this gene will have altered phytohormone ratios and the ensuing phenotype is likely to be of interest.

Control of Plant Development by Phytohormone-Specific ß-Glucosidases. Maize kernels are a rich source of auxin conjugates. These compounds accumulate in the endosperm during seed maturation and are mobilized to other parts of the seedling during germination (*23,24*). Bandurski and coworkers have already suggested that hydrolysis and transport of auxin conjugates from the endosperm to the shoot and to the root could be of importance for controlling maize seedling development.

p60 could play a pivotal role in the germination process by controlling the release of free auxin but, equally well, p60 could play a general role in plant growth control by specifically releasing active phytohormones from inactive conjugates in various tissues.

During evolution several soil bacteria have developed strategies allowing them to reprogramme plant physiological processes for their own benefit by altering phytohormone levels. For example, in addition to containing genes coding for enzymes involved in phytohormone synthesis, *A. rhizogenes* contains genes which are responsible for neoplastic root growth (*root loci, rol*). When expressed in transgenic tobacco plants two of these loci, *rolB* and *rolC*, were shown to drastically affect growth and development causing phenotypes reminiscent of those for altered phytohormone concentrations (*25*). The *rolC* gene, for example, induced phenotypes including reduced fertility, stimulation of cell division, inhibition of cell elongation and plant dwarfing. These phenotypyes are characteristic of alterations of cytokinin concentrations (*20*). Although the cell-autonomous action of the *rolC* gene did not suggest that the *rolC* product would play a role in phytohormone production (*26*), it turned out that the *rolC* gene encodes a ß-glucosidase capable of liberating free cytokinin from cytokinin-O- and -N-glucosides (*2*). Similarly, it was found that the product of the *rolB* gene can hydrolyze indoxyl-glucosides *in vitro*, suggesting that its effect would result from some alteration in the intracellular pool of active auxins (*3*).

An attractive model explaining plant growth control could be based on the action of ß-glucosidases. Transcriptional regulation of their genes could be linked to environmental cues and thus allow various developmental adaptations. Auxin and cytokinin conjugates have been found to be broadly distributed in plants. Thus, phytohormone-specific ß-glucosidases might represent a link between environmental stimuli and the release of active phytohormones from precise locations in the plant. Although these ideas are far from being proven, they open a promising area of research in plant development. We hope that future investigations will contribute to defining the importance of phytohormone-specific ß-glucosidases in the control of plant development.

Perspective

Bacteria like *A. rhizogenes* have evolved the capacity to use endogenous phytohormone pools by releasing active phytohormones with the help of specific ß-glycosidases. It is conceivable that plant genes encoding glycosidases or other hydrolases that specifically cleave phytohormone conjugates will influence the relative concentrations of active free and inactive bound phytohormones. Regulation of the genes encoding such hydrolytic enzymes by environmental or other plant-specific signals may be one of the mechanisms controlling plastic growth responses of plants. Genes like *Zmp60.1* or other members of this family from maize and arabidopsis will provide the tools to test this hypothesis.

Acknowledgements

Our work was supported by the Human Science Frontier Program Organisation and EEC BRIDGE contract CT-BIOT-90-0178.

Literature Cited

1.   Davies, P.J. *Plant hormones and their role in plant growth and development.* Martinus Nijhoff Publishers, Kluwer Academic Publishers Group, Dordrecht, The Netherlands, **1987**

2.   Estruch, J.J., Chrisqui, D., Grossman, K., and Schell, J. *EMBO J.* **1991**, *10*, 2889-2895.

3.   Estruch, J.J., Schell, J., and Spena, A. *EMBO J.* **1991**, *10*, 3125-3128.

4.   Schliemann, W. *Naturwissenschaften* **1991**, *78*, 392-401.

5.   Nowacki, J., and Bandurski, R.S. *Plant Physiol.* **1980**, *65*, 422-427.

6.   Pengelly, W.L., Hall, P.J., Schulze, A., and Bandurski, R.S. *Plant Physiol.* **1982**, *69*, 1304-1307.

7.   Chowdry, V., Westheimer, F.H. *Ann. Rev. Biochem.* **1979**, *48*, 293-325.

8.   Melhado, L.L., Jones, A.M., Leonhard, N.J., and Vanderhoef, L.N. *Plant Physiol.* **1981**, *68*, 469-475.

9.   Jones, A.M., Melhado, L.L., Ho, T.-H.D., Pearce, C.J., and Leonhard, N.J. *Plant Physiol.* **1984**, *75*, 1111-1116.

10.   Hicks, G.R., Rayle, D.L., Jones, A.M., and Lomax, T.L. *Proc. Natl. Acad. Sci. USA* **1989**, *86*, 4918-4952.

11.   Hicks, G.R., Rayle, D.L., Lomax, T. *Science* **1989**, *245*, 52-54.

12.   Campos, N., Feldwisch, J., Zettl, R., Boland, W., Schell, J., and Palme, K. *Technique* **1991**, *3*, 69-75.

13.   Campos N., Bako L., Feldwisch J., Schell J. and Palme K. *Plant J.* **1992**, *2*, 675-684.

14.   Stuber, C.V., Goodman, M.M., and Johnson, F.M. *Biochem. Genet.* **1977**, *15*, 383-394.

15.   Esen, A., and Cokmus, C. *Biochem. Genet.* **1990**, *28*, 319-336.

16.   Feldwisch J., Zettl R., Hesse F., Schell J. and Palme K. *Proc. Natl. Acad. Sci. USA.* **1992**, *89*, 475-479.

17.   Pryde J.G. *Trends in Biochem. Sci.* **1986**, *11*, 160-163.

18. Shimomura, S., Sotobayashi, T., Futai, M., Fukui, T. *J. Biochem.* **1986**, *99*, 1513-1524.
19. Hesse, T., Feldwisch, J., Balshüsemann, D., Bauw, G., Puype, M., Vandekerckhove, J., Löbler, M., Klämbt, D., Schell, J., and Palme, K. *EMBO J.* **1989**, *8*, 2453-2461.
20. Palme, K., and Schell, J. *Current Biology* **1991**, *2*, 228-230.
21. Tanimoto, E., and Pilet, P.-E. *Planta* **1978**, *138*, 119-122.
22. Love D.R., Fisher R. and Bergquist P.L. *Mol. Gen. Genet.* **1988**, *213*, 84-92.
23. Epstein E., Cohen J.D. and Bandurski R.S. *Plant Physiol.* **1980**, *65*, 415-421.
24. Cohen J.D. and Bandurski R.S. *Ann. Rev. Plant Physiol.* **1982**, *33*, 403-430.
25. Schmülling, T.J., and Spena, A. *EMBO J.* **1988**, *7*, 2621-2629.
26. Spena, A., Aalen, R.B., and Schulze, S.C. *Plant Cell* **1989**, *1*, 1157-1164.

RECEIVED April 9, 1993

Chapter 15

Stability and Activity of Plant and Fungal β-Glucosidases under Denaturing Conditions

Asim Esen and Gunay Gungor

Department of Biology, Virginia Polytechnic Institute and State University, Blacksburg, VA 24061

The stability and and the activity of maize β-glucosidase were studied in sodium dodecyl sulfate (SDS, 0.05 to 3.2%), sodium deoxycholate (DOC, 0.05 to 3.2%), N,N-dimethylformamide (DMF, 2.5 to 30%), urea (0.25 to 5 M), methanol (2.5 to 30%), and ethylene glycol (EG, 2.5% to 30%). The short-term stability and activity of almond β-glucosidase and the activity of two fungal (*Trichoderma and Penicillium*) were also studied in some of the same denaturants. Maize β-glucosidase retain about 50 to 70% of its initial activity up to 2 days and 20% of its initial activity up to 16 days in the presence of 0.1 to 3.2% SDS or DOC at 4°C. In contrast, it is inactivated completely in the presence of 2 to 5 M urea or 10 to 30% DMF after 2 days. The course and extent of inactivation is influenced by the concentration of denaturant, length of exposure to denaturant, and pH of denaturant medium. When enzyme activity is assayed in the above-mentioned denaturants, the activity decreases as the concentration of denaturant increases. The greatest decrease ($>95\%$) in activity was in urea and the lowest decrease (about 50%) in methanol. In general, maize β-glucosidase shows higher relative activity in all denaturants than almond β-glucosidase especially at higher denaturant concentrations. β-glucosidases from two fungal sources (*Trichoderma and Penicillium*) also show higher activity in denaturants than those from plant sources (i.e.,maize and almond). Furthermore, zymograms of both plant and fungal β-glucosidases can be developed in SDS, native, and isoelectric focusing gels after samples are treated with up to 3.2% SDS.

Studying the stability and activity of enzymes under conditions (e.g., extremes of pH and temperature, organic solvents, chaotropic agents, and detergents) that denature and inactivate typical proteins has potential for providing new approaches not only for purifying and characterizing proteins but also developing novel or unconventional conditions to carry

0097–6156/93/0533–0214$07.50/0
© 1993 American Chemical Society

out enzyme-catalyzed reactions in laboratory or industrial scale. With these considerations in mind, we have investigated the stability and activity of selected plant and fungal β-glucosidases in the presence of ionic detergents (sodium dodecyl sulfate and deoxycholate), chaotropic agents (urea) and organic solvents (dimethylformamide, ethylene glycol, and methanol). Here we define the "stability" in terms of the residual activity remaining after the enzyme has been incubated in a denaturant for a given length of time and then assayed under standard conditions. Similarly, we define the "activity" in terms of the residual activity measured when the enzyme is assayed in the presence of a denaturing agent. Only the maize enzyme was subjected to treatments with all denaturants. The stability and activity of other β-glucosidases were compared with that of the maize enzyme under some of the conditions used and defined for the maize enzyme.

Experience and conventional wisdom would dictate that the biological activity of proteins (e.g., catalytic activity of enzymes) would be lost, often irreversibly, in the presence of or after exposure to known protein denaturants. Therefore, one avoids denaturing conditions during purification and storage if the protein of interest is to be used in biological or biochemical activity assays. Thus investigation of the stability and activity of enzymes in the presence of denaturants has not received much attention. This is in spite of fact that retention and/or recovery of activity after treatment with a known denaturant such as SDS can provide considerable information about protein structure and function especially in the case of enzymes with quaternary structure, e.g., identification of subunits or monomers with catalytic activity, renaturation in situ after diluting or removing the denaturant, and prediction about the compactness and rigidity of the tertiary structure if resistance to denaturation is shown.

Enzyme stability and activity in the presence of denaturants have been largely studied on enzymes from thermophilic bacteria. These studies show that the enzymes caldolysin, α-glucosidase, glyceraldehyde 3-P dehydrogenase and malic enzyme isolated from thermophilic bacteria were fully stable in the presence of 1.0, 0.06, 0.1, and 0.05% SDS, respectively, for 10 min to 24 hr depending on the enzyme (*1, 2, 3, 4*). However, the stability was drastically reduced when the same enzymes were treated with SDS at or near optimum temperatures (60-75°C). For example, caldolysin was fully stable after exposure to 1.0% SDS at 18°C for 24 hrs but lost 50% of its activity at 75°C in 5 hrs (*3*). It is not surprising that a mesophilic enzyme, rabbit glyceraldehyde 3-P dehydrogenase, was completely inactivated by 0.1% SDS at 30°C in 10 min while its thermophilic counterpart from *Thermus thermophilus* retained full activity under the same conditions (*2*). In a few investigations the enzyme activity was measured in assay mixtures containing varying amounts of SDS. Yokoyama et al. (*5*) observed no inhibition of carboxypeptidase activity when SDS was present in the assay mixture at or below 0.1% but 57% and 95% inhibition were observed at concentrations of 0.5 and 1.0% SDS, respectively. Studies with malic enzyme and glyceraldehyde 3-P dehydrogenase also indicated that the concentrations of SDS that reduced stability also reduced activity when SDS was present in the reaction mixture, and again dramatic differences between mesophilic and thermophilic enzymes were noted (2, 4). Two other ionic detergents, cholate and its deoxy derivative (deoxycholate), were also studied for their effects on enzyme activity after their addition to assay mixtures. In the case of ATPase from a thermophilic bacterium cholate and DOC caused a ten-fold increase in enzyme activity at concentrations of 0.6 and 0.14%, respectively, but 50% inhibition was observed when beef heart and yeast ATPases were assayed in the presence of 0.05% and 0.5% DOC concentrations, respectively (6). In another study complete inhibition of rabbit ATPase and cAMP-hydrolase by 0.5% DOC was observed (7). In contrast, no inhibitory effect of DOC was found on malic enzyme from a thermophilic bacterium at and below 1.5% DOC in the reaction medium (4).

A review of the literature indicates that zymograms of a number of enzymes can be developed after SDS-PAGE. For example, Rosenthal and Lacks (8) showed that activities of various nucleases could be detected on SDS gels upon the removal of SDS by washing the gel after electrophoresis. Lacks and Springhorn (9) applied zymogram techniques to proteases, amylases and dehydrogenases after SDS-PAGE. Their data indicated that about 1-20% of the activity was recovered after SDS removal, and this level of recovery was sufficient to visualize the zones of activity in gels. Subsequently, a variety of other enzymes such as DNA polymerases (10), plasminogen activator (11), β-lactamase (12), protein kinases (13), tyrosinase (14), extracellular protease (15), phosphorylase kinase (16), lysozyme (17), phenoloxidase (18), and β-glucosidase (19) were shown to be amenable to detection on SDS gels. In most of these studies samples were not boiled to avoid complete denaturation and dissociation while in other studies they were boiled for a few minutes prior to electrophoresis (9, 12).

The effects of chaotropic agents such as urea and guanidine-HCl on enzyme stability and activity were investigated using enzymes both from mesophilic and thermophilic organisms. As was the case with ionic detergents, thermophilic enzymes were found to be more stable and active than their mesophilic counterparts after exposure to a chaotropic agent or when such an agent was present in the assay mixture. Only some representative cases will be mentioned here. A comparison of the enzymes 6-phosphogluconate dehydrogenase and malate dehydrogenase from *E. coli* with those from *Bacillus stearothermohilus* revealed that the *Bacillus* enzymes were fully stable in 4-8 M urea, which completely inactivated the *E. coli* enzymes under the same conditions (20, 21). Differences of the same magnitude were observed also between yeast and *Thermus* glyceraldehyde 3-P dehydrogenases (2). *Bacillus* α-glucosidase was another enzyme with high stability in 7-8 M urea in which *Thermus* caldolysin was inactivated after the same length of exposures (1, 3). In all cases, enzyme stability in urea decreased with increased urea concentration, temperature, and length of incubation.

Guanidine-HCl was found to have a more inactivating and inhibitory effect on enzymes than urea at comparable molar concentrations. For example, *Bacillus* 6-phosphogluconate dehydrogenase was completely stable in the presence of up to 8 M urea while it was completely inactivated in 4.5 M guanidine-HCl (*22*). However, some enzymes such as caldolysin exhibited essentially the same level of stability in the presence of urea and guanidine-HCl (*3*).

There are also many reports on enzyme stability and activity in the presence of organic solvents as well as enhancement of activity after exposure to certain organic solvents. Some examples of the enzymes investigated in this respect were yeast and *Bacillus* 6-phosphogluconate dehydrogenases (*22*), *Bacillus* α-glucosidase (*1*), Jackbean urease (*23*), and *Sulfolobus* malic enzyme (*4*). For example, *Bacillus* 6-phosphogluconate dehydrogenase showed little or no activity loss when treated with concentrations of acetone, diaxon, and DMF below 40% while yeast phosphogluconate dehydrogenase was completely inactivated by 20% acetone or diaxon at 20°C after 90 min exposure (*1*). In the case of *Sulfolobus* malic enzyme no inactivation by 50% DMF or methanol was observed after 24 hrs of treatment but ethanol at the same concentration led to 80% activity loss. When methanol and ethanol were added to the assay mixture at 50% final concentrations they caused 70-100% enhancement of activity while acetone, 2-propanol and DMF caused only moderate to little enhancement at the same concentration. These and other studies suggested that stability and activity of enzymes in the presence of organic solvents were correlated with hydrophobicity of the organic solvent in that higher hydrophobicities were less detrimental to activity and stability than low hydrophobicities (*24*). In this connection, performing enzymatic reactions and processes in organic solvents have been an area of some investigation (see review by *25*). For example, studies in the laboratory of M. Klibanov show that the proteolytic enzymes subtilisin and α-chymotrypsin can act as catalysts in dry organic solvents such as hexadecane, octane, and others (*26*). The rationale for these studies is to develop alternatives to traditional enzyme purification and characterization approaches in aqueous solvents as well as to develop industrial scale enzymatic processes in nonaqueous solvents.

During our characterization studies on maize β-glucosidase, we observed serendipitously that the enzyme maintained much of its catalytic activity both after prolonged incubation with SDS and when SDS was present in the reaction medium. We also showed that the activity of β-glucosidases from plant and fungal sources can be detected on SDS gels as well as on non-denaturing gels after samples are prepared and applied in the presence of up to 3.2% SDS (*19*). Based on these results, we performed a systematic study of β-glucosidase stability and activity in the presence of SDS and other protein denaturants. Table I lists the denaturants and their concentrations used in these studies. We show that maize, almond, *Penicillium* and *Trichoderma* β-glucosidases exhibit varying degrees of activity and stability in the presence of the denaturants listed in Table I.

Materials and Methods

Materials. Crude and purified fractions of maize β-glucosidase were

Table I. Denaturants and Their Concentrations Tested For Effects on
β-Glucosidase Stability and Activity

Denaturant	Concentration
Anionic detergents	
Sodium Dodecyl sulfate	0.05-3.2%
Deoxycholate (DOC)	0.05-3.2%
Chaotropic agents	
Urea	0.5-6 M
Organic solvents	
Dimethylformamide (DMF)	2.5-30%
Methanol (MeOH)	2.5-30%
Ethylene glycol (EG)	2.5-30%

prepared from the maize inbred K55 in our laboratory (27). Purified and partially purified preparations of Emulsin (almond β-glucosidase) and cellulase complexes of *Trichoderma viride* and *Penicillium funiculosum*, (both of which contain β-glucosidase) were obtained from Sigma.

Enzyme Extraction and Preparation. Shoots (mesocotyl, apex, primordial leaves, and coleoptile) from 5-6 day-old maize seedlings were harvested, ground in an ice-chilled mortar with a pestle, and extracted with two different buffers. The buffers were: 100 mM Tris-HCl, pH 8.0 and 50 mM Na-Acetate, pH 5.0. The ratio of buffer volume (ml) to fresh weight (g) was 3:1. The homogenates were centrifuged at 17,000 X g for 15 min. The supernatant was used in stability and activity assays. In addition, maize β-glucosidase was purified by a combination of differential solubility and chromatography as described (27). Two homogeneous fractions so obtained were used for stability and activity assays in the presence of denaturants as well as for zymogram assays.

Enzyme Assays. β-glucosidase activity in crude extracts or purified preparations was spectrophotometrically measured using the chromogenic substrate p-nitrophenyl-β-D-glucose (pNPG). Routinely, 15 μl of the enzyme (the soluble extract) were diluted with 1485 μl of 10 mM citrate-20 mM phosphate buffer, pH 5.5, in a 1.5 ml microfuge tube. Seventy μl of diluted enzyme were incubated in the wells of 96-well microtiter plates with 70 μl of 8 mM pNPG in the same buffer at 25° C for 5 min. The reaction was stopped by the addition of 70 μl 400 mM sodium carbonate, and the p-nitrophenol liberated from pNPG was measured at 410 nm.

Dependence of Stability in the Presence of Denaturants on pH. In order to determine the effect of pH on enzyme stability in the presence of SDS, pure maize β-glucosidase powder was dissolved in 0.1 M Tris (pHs 10, 9, 8), 0.1 M phosphate (pHs 7, 6), and 0.1 M acetate (pHs 5, 4, 3). One pH treatment set contained SDS at a final concentration of 0.1% (w/v) while the other set lacked SDS (control set). Each set was divided into two subsets, one of which was incubated at 4°C for 0.5, 1 and 2 hrs, and 1, 3 and 8 days while the other was incubated at room temperature (approx. 23-25 °C) for 4 hrs, 1, 3 and 8 days. At the end of each incubation period, aliquots were taken, diluted 100-fold into the assay buffer (10 mM citrate-20 mM phosphate, pH 5.5), and assayed under standard conditions. Residual activity remaining after exposure to SDS at each pH, temperature, and time of incubation was expressed as percent of controls for each treatment.

Stability of Maize β-Glucosidases in the Presence of SDS, DOC, Urea, and DMF. The purified enzyme was prepared in 0.1 M Tris-HCl, pH 8.0 and aliquoted to microfuge tubes, and the denaturant was added to tubes so that SDS and DOC would increase serially from 0.05 to 3.2% (w/v), urea from 0.5 to 6 M, and DMF from 2.5 to 30% (v/v) final concentration. Tube contents were divided to seven microfuge tubes and placed at 4°C. A tube from each denaturant treatment was removed and assayed at day 0, 1, 2, 4, 8, and 16 after diluting 100-fold in the assay buffer. Residual activity after each time of incubation in the presence of each denaturant was expressed as percent of control (untreated).

Comparison of the Stability and Activities of Maize and Almond β-Glucosidases in Denaturants. Maize and almond β-glucosidase activities were assayed in the presence of 0.016 to 2.0% SDS, 0.25 to 5 M urea, and 2.5 to 30% MeOH. Activity data were presented as percent of control. In addition, both maize and almond β-glucosidases were prepared and incubated for 3 and 30 min, respectively, in the presence of 0.025 to 3.2% SDS in buffers of two different pHs, 5 and 8 at 25°C. The purpose of these experiments was to study the effects of pH, time of incubation, and SDS concentration on the stability of enzymes over a short time course at 25 °C.. Residual enzyme activity was expressed as percent of control.

Activity of Two Fungal β-Glucosidases in SDS, Urea, DMF, and EG. Maize, *Trichoderma*, and *Penicillium* β-glucosidases were dissolved in water and diluted with 100 mM acetate, pH 5.0 in a 1:1 ratio to yield approximately one A410 unit of p-nitrophenol (pNP) when assayed under standard conditions. These enzyme solutions were aliquoted to microfuge tubes, and the denaturant was added to tubes so that SDS would increase serially from 0.025 to 3.2%, urea from 0.25 to 5.0 M, DMF from 2.5 to 30%, and EG from 2.5 to 30% final concentration. Tube contents were mixed, incubated on ice for 10 min, and then assayed under standard conditions.

Electrophoretic Methods. Maize β-glucosidase (crude and purified) was prepared in 50 mM NaAc, pH 5.0 and 100 mM Tris-HCl, pH 8.0 to yield approximately one A410 unit of p-nitrophenol (pNP) when assayed under standard conditions. These enzyme solutions were aliquoted to microfuge tubes, and SDS was added to tubes so that it would increase serially and be 0.025, 0.05, 0.1, 0.2, 0.4, 0.8, 1.6 and 3.2%. Glycerol was added to all samples at a final concentration of 10% before electrophoresis. SDS-treated and control (untreated) samples were then brought to room temp for 10 min and applied to the gel. In addition, purified maize β-glucosidase was prepared in 50 mM Tris-HCl, pH 6.8, containing 1% SDS plus either 5% 2-ME or no 2-ME. The 2-ME treated sample was applied both before after boiling while the one that was not treated with 2-ME was applied without boiling. β-glucosidase activity on gels was detected following SDS-PAGE, native PAGE, and IEF (*19*). Almond, *Trichoderma*, and *Penicillium* β-glucosidases were also subjected to to the same electrophoretic analyses as those used for maize β-glucosidase (*19*). All electrophoretic fractionations were performed at 5°C.

Results

Dependence of Enzyme Stability on pH in SDS. Stability of maize β-glucosidase in SDS was dependent on pH based on time-course studies at 4°C extending over 8 days (Figure 1). The enzyme was rapidly inactivated at and below pH 4 but exhibited the same stability in the pH range 5-10 in the first hour of exposure to SDS (Figure 1). However, up to 30% activity loss occurred at pH 5, 6 and 10 as the length of incubation increased from 2 hours to 3 days. At day 8, the samples incubated at pH 5 and 6 were found to have lost 65% and 40% of their activity, respectively, while there was little or no additional activity loss in samples incubated in the pH range 7-10 between 3 and 8 days. Thus the data in Figure 1 show that the enzyme is most stable between pH 7 and 9, and activity loss in this range was less than 10% when compared to control (w/o SDS) samples subjected to the same pH treatments.

Stability of Maize β-Glucosidase in the Presence of SDS, DOC, Urea, and DMF. Stability of maize β-glucosidase in ionic detergents (SDS and DOC), urea, and organic solvents were followed over a period of 16 days. The data in Figure 2 show that only 2 to 18% reduction in activity occurred in the presence of 0.05 to 3.2% SDS in the first 24 hours of incubation. At day 2 the greatest activity loss was 35 to 45% while at day 4 it was 60 to 65% at and above 0.1% SDS concentrations. Activity loss reached 75-80% of control samples at and above 0.1% SDS by day 8, and no further activity losses occurred between days 8 and 16. Incubation of the enzyme in the presence of DOC under conditions identical to those used for SDS treatments yielded essentially the same results, as shown in Figure 3. The only difference was additional activity losses in the presence of 0.2-0.4% DOC between days 8 and 16.

The stability data (Figure 4) indicate that maize β-glucosidase is less stable in urea than in SDS and DOC. For example, enzyme activity is

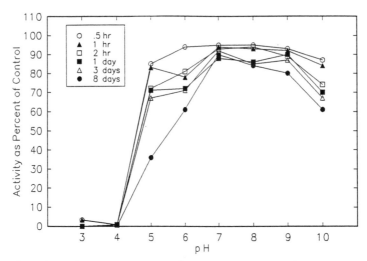

Figure 1. Stability of maize β-glucosidase in SDS in the pH range 3 to 10. Control treatment did not include SDS.

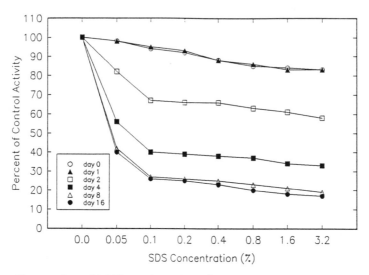

Figure 2. Stability of maize β-glucosidase in sodium dodecyl sulfate (SDS) at pH 8 at 4°C. (Spacing on the X axis is not linear in order to show the trend at lower SDS concentrations).

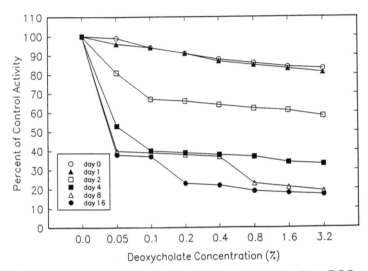

Figure 3. Stability of maize β-glucosidase in sodium DOC at pH 8 at 4°C. (Spacing on the X axis is not linear in order to show the trend at lower DOC concentrations).

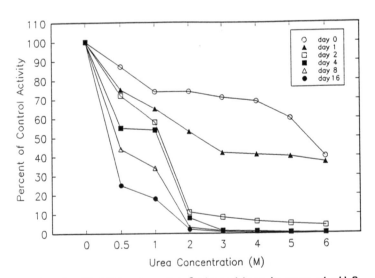

Figure 4. Stability of maize β-glucosidase in urea at pH 8 at 4°C. (Spacing on the X axis is not linear in order to show the trend at lower urea concentrations).

rapidly lost decreasing from 87% of the control in 0.5 M urea to 40% in 6 M urea within 10 minutes. There was little or no activity remaining in samples exposed 2 to 6 M urea for one day or longer, and the activity gradually decreased to 15 to 20% of control in the presence of 0.5 to 1 M urea by day 16. Both the time-course and the extent of the activity changes in the presence of 2.5% (0.32 M) to 30% (3.84 M) DMF showed considerable similarity to those of urea, the major difference being that the rate of activity loss was much slower in the first day of incubation (Figure 5). At day 2 and after there was little or no measurable activity in samples incubated in 10-30% DMF, and the residual activity in samples incubated in 2.5 and 5% DMF were, respectively, 28 and 20% of control.

Sort-Term Stability of Maize and Almond β-Glucosidases in SDS. Activities of maize and almond β-glucosidases decreased after exposure to SDS as the length of incubation time and/or SDS concentration increased and pH decreased. The data show that there was little or no change in enzyme activity at pH 8.0 after 30 min of exposure up to 0.4% final SDS concentration (Figure 6). At the highest SDS concentrations (1.6 and 3.2%), the almond enzyme was more stable (only 5% activity loss) than the maize enzyme (40-50% activity loss). In contrast, the maize enzyme was clearly more stable than the almond enzyme at pH 5.0 under the same conditions. The difference between the maize and almond enzymes in stability in the presence of SDS at pH 5 is most apparent after exposure to 0.1% SDS for 30 min during which the former retained about 90% of its original activity while the latter lost all of its activity when compared to untreated (control) samples (Figure 7). The data indicated that the almond enzyme was more stable at or above 0.8% SDS at pH 8.0 than the maize enzyme, but the maize enzyme was more stable at or above 0.05% SDS at pH 5.0 than the almond enzyme. In addition, the almond enzyme was inactivated completely within 3 minutes at and above 0.1% SDS concentrations at pH 5.

Activity of Maize and Almond β-Glucosidases in Denaturants. When SDS was added to the assay mixture at concentrations varying from 0.006 to 2.0% the activity of maize β-glucosidase decreased linearly to about 45% of control activity as SDS concentration increased from 0.006 to 0.1%. There was no additional activity decrease above 0.1% SDS, that is between 0.1 and 2.0% SDS (Figure 8). When the effects of urea (0.25 to 5.0 M) and SDS (0.006 to 2.0%) in the assay mixture were compared, the inhibitory effect of urea increased in proportion to its concentration approaching 100% at 5 M (Figure 9). In contrast, the inhibitory effect of SDS stabilized around 60% in 0.05 to 0.1% SDS, and there was no further inhibition above 0.1%.

SDS added to the reaction medium was more inhibitory to the almond enzyme at higher concentrations (0.1% or higher) than the maize enzyme as shown in Figure 8. In contrast, SDS was more inhibitory to the maize enzyme at lower (<0.05%) concentrations than the almond enzyme. Urea inhibited the activity of both maize and almond β-glucosidase in a concentration dependent manner; inhibition was near 100% in the

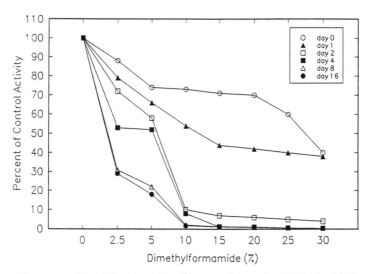

Figure 5. Stability of maize β-glucosidase in DMF at pH 8 at 4°C. (Spacing on the X axis is not linear in order to show the trend at lower DMF concentrations).

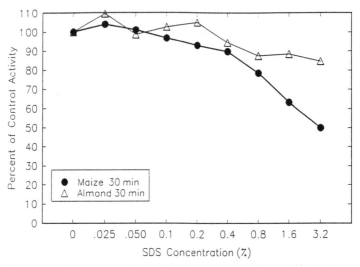

Figure 6. Short-term (30 min) stability of maize and almond β-glucosidases in SDS at pH 8 at 25°C. (Spacing on the X axis is not linear in order to show the trend at lower SDS concentrations).

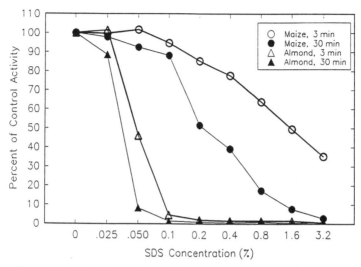

Figure 7. Short-term (3 and 30 min) stability of maize and almond β-glucosidases in SDS at pH 5 at 25°C. (Spacing on the X axis is not linear in order to show the trend at lower SDS concentrations).

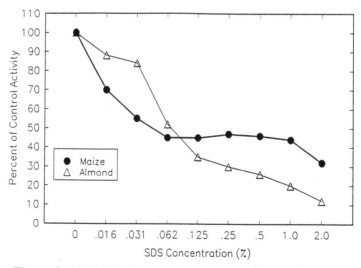

Figure 8. Activity of maize and almond β-glucosidases in SDS. (Spacing on the X axis is not linear in order to show the trend at lower SDS concentrations).

presence of 5 M urea and 50% in the presence of 0.5 to 1.0 M urea. However, urea was slightly less inhibitory to the maize enzyme than to the almond enzyme at all concentrations (Figure 9).

When the activities of maize and almond β-glucosidases were assayed in methanol concentrations of 2.5 to 30%, lower concentrations of methanol (2.5-10%) resulted in a slight enhancement of activity in the case of the maize enzyme, while higher concentrations (20-30%) were slightly inhibitory. In contrast, there was no enhancing effect of methanol on almond β-glucosidase, and methanol inhibited activity at all concentrations between 5 and 30%. Highest inhibition of maize β-glucosidase activity by methanol was 30% and that of almond β-glucosidase was 60%, both occurring in 30% methanol (Figure 10).

Activity of Fungal Enzymes in Denaturants. Figure 11 shows the activity profiles of maize, *Trichoderma* and *Penicillium* β-glucosidases. These data clearly show that *Trichoderma* β-glucosidase was the most resistant to inhibition by SDS, where the inhibition even at the highest concentration of SDS (3.2%) was about 10%. *Penicillium* β-glucosidase was the second best in terms of activity in SDS where the activity decreased gradually to 55% of control (45% inhibition) as SDS concentration increased to 3.2%. Although the activity profiles of the *Penicillium* and maize enzymes were similar at and below 0.2% SDS, concentrations of SDS higher than 0.2% were significantly more inhibitory to the maize enzyme and reached about 70% in 0.8-3.2% SDS.

When the activities of the three enzymes were compared in the presence of urea, they were all inhibited by this chaotropic agent, inhibitory effect increasing with urea concentration. However, relatively speaking, the inhibitory effect of urea was less on two fungal β-glucosidases than the maize enzyme (Figure 12).

The activity profiles (Figure 13) derived from assays of the three enzymes in the presence of DMF concentrations varying from 2.5 to 30% were almost identical to those obtained from assays in the presence of urea (Figure 12).

Finally, activities of the three enzymes were compared in the presence of EG concentrations varying from 2.5 to 30%. Activity profiles (Figure 14) show that EG inhibited the maize enzyme most followed by the *Penicillium* and *Trichoderma* enzymes. At the highest concentrations of EG (30%), *Trichoderma*, *Penicillium* and maize β-glucosidases were inhibited, respectively, 25, 60, and 80%.

SDS-PAGE Zymograms of Maize β-Glucosidase. Figure 15 shows that one can obtain zymograms of maize β-glucosidase after adding SDS from 0.025 to 3.2% final concentration to crude or purified enzyme preparations. β-glucosidase activity became visible as an intense band(s) in the 110-120 kD region of the gel with no apparent decrease in band intensity of extracts made with a pH 8 buffer (Tris-HCl) as SDS concentration increased, while nearly 50% decrease in band intensity was observed in extracts made with pH 5 buffer (acetate) at or above 1.6% SDS concentration.

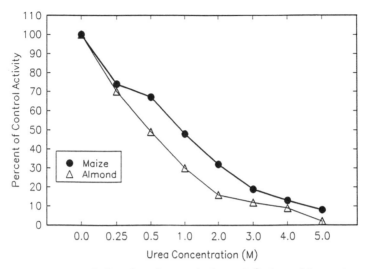

Figure 9. Activity of maize and almond β-glucosidases in urea. (Spacing on the X axis is not linear in order to show the trend at lower urea concentrations).

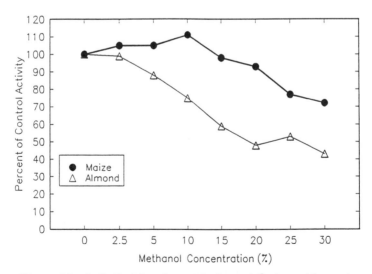

Figure 10. Activity of maize and almond β-glucosidases in methanol. (Spacing on the X axis is not linear in order to show the trend at lower methanol concentrations).

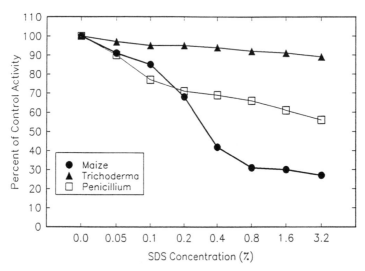

Figure 11. Activity of *Trichoderma, Penicillium*, and maize β-glucosidases in SDS. (Spacing on the X axis is not linear in order to show the trend at lower SDS concentrations).

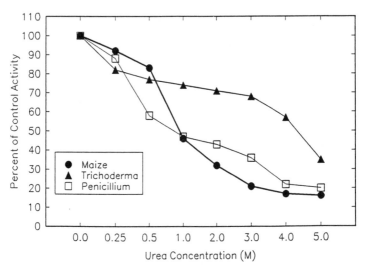

Figure 12. Activity of *Trichoderma, penicillium*, and maize β-glucosidases in urea. (Spacing on the X axis is not linear in order to show the trend at lower SDS concentrations).

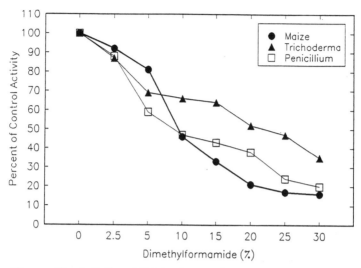

Figure 13. Activity of *Trichoderma, Penicillium,* and maize β-glucosidases in DMF. (Spacing on the X axis is not linear in order to show the trend at lower DMF concentrations).

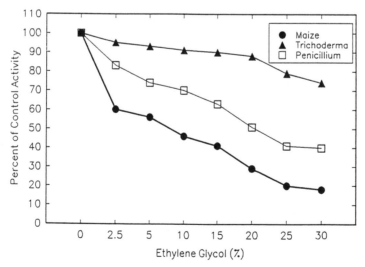

Figure 14. Activity of *Trichoderma, Penicillium,* and maize β-glucosidases in EG. (Spacing on the X axis is not linear in order to show the trend at lower EG concentrations).

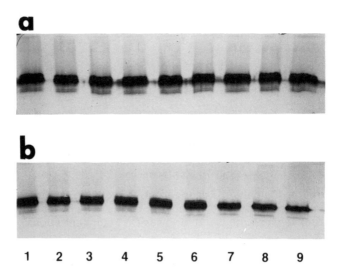

Figure 15. β-Glucosidase zymograms of maize coleoptile extracts after treatment with SDS and electrophoresis through a 10% SDS gel (anode at bottom). **a**, 100 mM Tris-HCl, pH 8.0 extract. **b**, 50 mM NaAc, pH 5.0 extract. Extractions were made with the aforementioned buffers, SDS was added to samples of 0.025 to 3.2% final concentrations, samples were electrophoresed, and the gels were stained for β-glucosidase activity. Lane 1, control (no SDS added). Lanes 2-9, after adding SDS to a final concentration of 0.025 to 3.2%. 2, 0.025%; 3, 0.05%; 4, 0.1%; 5, 0.2%; 6, 0.4%; 7, 0.8%; 8, 1.6%; 9, 3.2%.

When crude enzyme samples were subjected to 0.31 to 5% 2-ME in the presence of 2% SDS, there was no visible effect of 2-ME on maize β-glucosidase activity profiles (*19*). However, when purified enzyme fractions were used, a band of the same mobility as the 60 kD monomer appeared in addition to the 120 kD band in gels to which SDS and 2-ME treated samples were applied. Boiling led to irreversible activity loss irrespective of SDS and 2-ME concentration (Figure 16) but it resulted in total conversion of the 120 kD band (dimer) to the 60 kD band (monomer) as observed in Coomassie stained gels (Figure 16).

Native PAGE and IEF Zymograms of Maize and Other β-Glucosidase After SDS Treatment. β-glucosidase zymograms were also developed on native PAGE gels in the usual manner after SDS was added to samples (*19*).

When SDS-treated samples were fractionated by IEF and stained for activity, a doublet band in the pH 5.0 region of the gel was detected in both the control lane (no SDS) and those receiving the samples treated with 0.025 to 0.8% SDS. However, the banding pattern was distorted in lanes that received the samples containing 1.6% and 3.2% SDS (*19*).

Trichoderma and *Penicillium* and almond β-glucosidases also yielded zymogram patterns after SDS-PAGE when SDS was added to samples prepared in pH 8 buffer. In contrast, when SDS was added to samples prepared in pH 5 buffer, the almond enzyme activity zones were visible only in the control and in the sample treated with the lowest concentration (0.025%) of SDS, and none was visible at SDS concentrations of 0.1% or above (*19*).

Like the two higher plant β-glucosidases, fungal β-glucosidases also yielded bands with enzyme activity on SDS gels. In the case of *Penicillium* β-glucosidase, there were two zones of activity, a darker and slower band and a lighter but faster one. The *Trichoderma* profile also contained two zones of activity, a lighter one with slow mobility and a darker one with faster mobility (*19*).

Discussion

This study indicates that β-glucosidases from maize, almond, *Trichoderma and Penicillium* exhibit considerable stability and/or activity in the presence of common protein denaturing agents. This magnitude of relatively high stability and activity after exposure or in the presence of denaturants, although common in enzymes from thermophilic organisms, was not expected from the β-glucosidases, all from mesophilic organisms, included in our study. The data also clearly indicate that SDS treatment without heating does not fully inactivate maize, almond, *Trichoderma* or *Penicillium* β-glucosidases. This finding suggests that these enzymes have a relatively compact and rigid tertiary structure making their hydrophobic core inaccessible to SDS and other denaturants which disrupt or weaken hydrophobic interactions. The compactness and rigidity of tertiary and quaternary structures were thought to be responsible for the stability of thermophilic enzymes to denaturants especially at temperatures that lead to complete inactivation of mesophilic enzymes by the same denaturants

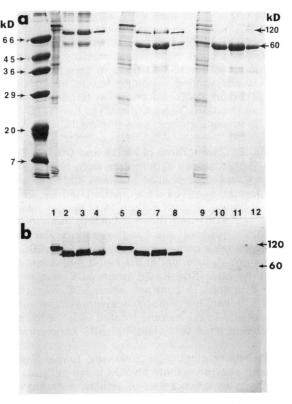

Figure 16.   Catalytically active form of maize β-glucosidase is a dimer.   SDS-PAGE (10% T, 4% C) showing profiles of coleoptile extract and Accell CM fractions (purified enzyme) after staining with Coomassie blue for general protein (**a**,) and with specific enzyme substrate for β-glucosidase activity (**b**,). Lane 1, 5, and 9, fresh coleoptile extract made with 50 mM NaAc, pH 5.0, containing 0.2 mg/ml TLCK (protease inhibitor); lanes 2-4, 6-8, and 10-12, three different pooled β-glucosidase fractions purified by Accell CM chromatography. Purified fractions have higher anodal mobility due to the modification by an endogenous SH-proteinase. Lanes 1-4, samples prepared in 1% SDS; lanes 5-8, samples prepared in 1% SDS+2.5 % 2-ME; lanes 9-12, samples prepared in 1% SDS+2.5% 2-ME and boiled. Note that the catalytically active form of the enzyme has slower mobility (larger in size; approx. 120 kD) than the inactive monomer (approx. 60 kD). Note also the conversion of the dimer (active) to monomer (inactive) partially after treatment with 2-ME (lanes 5-8) and completely after boiling (lanes 9-12). Unmarked lane on the left contain molecular weight markers.

(*1, 6, 22*). The stability and activity data obtained with β-glucosidases suggest that these enzymes resemble thermophilic enzymes in having compact and rigid structure at and below room temperature. Is the residual activity measured after exposure to or in SDS and other denaturants due to partially denatured but still catalytically active enzyme, or is it due to a fraction of enzyme molecules which retain native conformation? This question cannot be answered because no physical measurements such as CD, ORD, and UV spectra were made on β-glucosidase samples after exposure to or in denaturants.

Stability of Maize and Almond β-Glucosidase in Denaturants. Stability of maize β-glucosidase in SDS and possibly in other denaturing milieu appeared to be strongly dependent upon pH based on the activity measurements after exposing the enzyme to 0.1% SDS in buffers of pH 3 to 10 (Figure 1). The enzyme was most susceptible to inactivation (i.e. denaturation) at acidic pH values, most notably pH 4, at which SDS-treated enzyme was completely inactivated while its untreated counterpart had 85% of the activity of that incubated at pH 8 in the absence of SDS (control treatment with the highest activity). Thus almost all of the activity loss at pH 4 is attributable to the effect of SDS while only 25% is attributable to SDS at pH 3.0. This is because the enzyme samples not treated with SDS lose about 75% of their activity after 30 min of exposure to pH 3 in comparison with that at pH 8.0. In fact, sensitivity to SDS denaturation is still apparent, although substantially reduced, at pH values 5 and 6. These results suggest that ionic interactions (i.e., salt bridges) play an important role in the conformational stability of maize β-glucosidase and such interactions are disrupted or weakened, leading to varying extents of unfolding, at lower pH values. These partially unfolded "borderline" intermediates would be accessible to SDS for disruption of hydrophobic interactions leading to further unfolding and complete inactivation.

In view of pH dependence of the stability in SDS, we chose two pHs to carry out stability studies: 8, at which the greatest stability was observed, and 5, which is near the pH optimum (pH 5.5-5.8) of the enzyme and is the one at which susceptibility to SDS inactivation begins. The data in Figure 2 clearly indicate that the stability of maize β-glucosidase to SDS at pH 8 is essentially independent of SDS concentration when exposure time is one day or less. However, there is definitely time-dependent activity loss in concentrations of SDS especially above 0.05%. The fact that there is little or no difference between 0.1% and 3.2% SDS concentrations in terms of their denaturing activity suggests that the amount of SDS present in 0.1% solution is sufficient to saturate all the sites of interaction in the enzyme, and thus SDS in excess of 0.1% is ineffective. Alternatively, the effective concentration of SDS monomers in all solutions above 0.1% SDS is similar to that in 0.1% solution because excess SDS form crystalline precipitates at 4°C at higher SDS concentrations. Indeed we observed visible precipitates of SDS at such concentrations. Since the stability profiles obtained by DOC (Figure 3) are similar to those obtained by SDS, it is safe to conclude that these two ionic detergents are equivalent in terms of their mode of interaction with β-glucosidase.

The extreme sensitivity of maize β-glucosidase and other β-glucosidases to urea inactivation especially at or above 2 molar was somewhat unexpected in view of their relatively high stability in ionic detergents. It is known that ionic detergents primarily disrupt hydrophobic interactions (6, 28). Urea is also thought to disrupt hydrophobic interactions (29, 30), and possibly hydrogen bonding. Thus both hydrophobic groups and hydrogen bonds would be more accessible to a small molecule like urea. It is not surprising that urea inactivation of maize β-glucosidase (active form is a dimer) occurred rapidly even at low (<2 M) molar concentrations. Moreover, urea is fully soluble and active at the temperature of incubation (4°). Another explanation for increased susceptibility to urea denaturation may be the increased susceptibility of the enzyme to endogeneous proteases present in samples. The similar sensitivity of maize β-glucosidase to inactivation by DMF and urea suggest at least in part a similar mode of denaturation by both denaturants. This is because DMF has substituents (CH_3, $-NH_2$) capable of disrupting both hydrogen bonding and hydrophobic interactions (2).

When maize (monocot) β-glucosidase is compared to almond (dicot) β-glucosidase, the two enzymes exhibit significant differences in stability in denaturants. It is clear that the almond enzyme is more stable than the maize enzyme in the presence of higher SDS concentrations at pH 8 for 30 min (Figure 6) while the opposite is found when incubation with SDS is at pH 5.0 (Figure 7). In fact, the inactivation of the almond enzyme is complete in 3 min (Figure 7) in the presence of 1.0% SDS and higher.

Activity of maize and other β-glucosidases in denaturants. The activity profile of maize β-glucosidase obtained in the presence of 0.006 to 2.0% SDS in the reaction mixture clearly show that SDS inhibits enzyme activity in a concentration-dependent manner (Figure 8). However, even at the highest concentration of SDS (2%) in the reaction mix inhibition is about 65% and thus not complete. If one compares the activity loss after brief exposure to SDS at pH 5 and the activity inhibition in the presence of corresponding SDS concentrations in the reaction mixture at a comparable pH (5.5), percent activity loss and percent inhibition values are essentially similar. Thus the lack of any clear-cut indication of renaturation suggest that inactivation by SDS is both rapid and irreversible around pH 5.0 under the conditions employed. It is conceivable that after the dilution or dialysis of SDS and other denaturants, the enzyme can renature slowly to a catalytically active conformation. We have not done a long-term and systematic study of β-glucosidase renaturation to check this possibility.

In contrast to SDS, the activity profile in the presence of urea indicates essentially a linear activity decrease from 100% to zero as the concentration of urea increases (Figure 9). Interestingly, one obtains the same activity profile whether enzyme samples are exposed briefly to a given urea concentration and then diluted 200x for activity assay or that same concentration of urea is added to the reaction medium. This would also suggest that urea inactivation is rapid and irreversible because otherwise an enzyme sample exposed to 5 M urea and then diluted 200x should have a much higher activity when assayed without any urea in the reaction mixture than the one assayed in the presence of 5 M urea.

When the activities of maize and almond β-glucosidases are compared in the presence of SDS, a much greater inhibitory effect on the activity of the almond enzyme was evident at higher SDS concentrations in the reaction medium (Figure 8). This was consistent with higher sensitivity of the almond enzyme to inactivation by SDS at lower pH values such as pH 5. In fact, the almond enzyme appeared to be less sensitive to SDS in the assay mixture (pH 5.5) than to that with which it was incubated prior to the assay (Figure 7). For example, a 3 min exposure to 0.1% SDS at pH 5 completely inactivated the almond enzyme while the presence of 0.5% SDS in the reaction mixture inhibited activity only 50%. The basis for this discrepancy is not known but it may result from the presence of the substrate in the assay mixture such that the binding of the substrate partially protects the enzyme from inactivation. This interpretation is consistent with the observation that the resistance of enzymes to proteases and stability in denaturants increases after ligand or substrate binding because such binding makes proteins more compact and rigid. The presence of urea in the reaction mixture was more or less equally inhibitory to both the maize and almond enzyme with no apparent protection from the presence of substrate in the reaction medium.

As for the effects of organic solvents on β-glucosidase activity, the presence of methanol in the assay mixture up to 30% revealed considerable differences between maize and almond β-glucosidases. These differences involved slight enhancement of maize β-glucosidase at lower alcohol concentration and its less inhibition than that of the almond enzyme at higher methanol concentrations (Figure 10). Enzyme activity in organic solvents and enhancement of activity especially low concentrations of certain organic solvents have been well-documented in the literature (2, 4, 22, 25). These limited studies with two plant β-glucosidases suggest that these enzymes may have substantial activity at higher concentrations of methanol than the 30% tested. It is thus worth investigating β-glucosidase activity in other alcohols over a wider concentration range. It is suggested that organic solvents loosens the secondary and tertiary structure by disrupting hydrogen bonding and hydrophobic interactions and thus resulting in a more relaxed and flexible conformation, which can activate enzymes having a compact and rigid structure, such as most thermophilic and some mesophilic enzymes (4, 22, 23). However, organic solvents, depending on their chain length, branching, and hydroxyl contents, were shown to denature proteins above a certain concentration (4, 31).

The results of activity assays in the presence of denaturants using *Trichoderma* and *Penicillium* β-glucosidases indicate that the unusual resistance of maize β-glucosidase to denaturants is shared by β-glucosidases from other sources. In fact, fungal β-glucosidases, especially that of *Trichoderma*, appear to be totally stable in the presence of 3% SDS or less (Figure 11). However, the two fungal enzymes, especially that of *Penicillium*, exhibit no clear-cut superiority to maize β-glucosidase in terms of inhibition by urea and DMF (Figures 12-13).

These results may mean that hydrogen bonding plays an important role in maintaining the native or functional structure of β-glucosidases. Inhibition of activity by varying urea concentrations has been observed with other enzymes from other organisms (see above). Also, EG added to the reaction mixture was not sufficient to inhibit the *Trichoderma* enzyme (Figure 14). This organic solvent is believed to disrupt hydrophobic interactions, suggesting that either the *Trichoderma* enzyme is not primarily stabilized by hydrophobic interaction or the interior of the molecule is not accessible to the denaturant because it has a more compact and rigid tertiary structure than other β-glucosidases.

Our electrophoretic data (*19*) show that β-glucosidase zymograms can be developed on SDS, native and IEF gels after applying samples prepared in solutions with varying concentrations of SDS. These data support the results from solution treatments and assays in the presence of SDS and other denaturants. The electrophoretic variants of maize β-glucosidase observed on native gel zymograms when samples were prepared in pH 5 buffer are judged to be artifactual, resulting from the action of an SH-proteinase present in extracts of the shoot tissue (*32*). In SDS gels, a minor band showing β-glucosidase activity in the 100 kD region of gels is observed (Figure 15), which may also be a product of proteolysis.

In the case of the maize enzyme, both the genetic data and the zymogram patterns obtained under non-denaturing conditions suggest that the native enzyme is a dimer (*33*). The activities of maize and other dimeric β-glucosidases after SDS treatment and on SDS gels suggest that either SDS fails to dissociate dimers or the monomers reassociate to dimers upon the removal of SDS during equilibration washes. If indeed the native enzyme is a dimer, the activity after SDS treatment or on SDS gels suggests that the enzyme dimers do not dissociate in SDS unless heat is employed. The fact that addition of a reducing agent (e.g., 2-ME up to 250 mM final concentration) to the incubation and reaction media cause only partial dissociation of dimers without heating the maize β-glucosidase indicates that disulfide bonds are buried and thus not readily amenable to reduction. It is then possible that either disulfide bonds link monomers or the reduction of intramolecular disulfide bonds increase the susceptibility of dimers to denaturation by SDS.

In the case of maize β-glucosidase, heating at 65°C or above irreversibly inactivates the enzyme (*19* and Figure 16). This is in contrast to recovery of activity after removal of SDS in such enzymes as proteases, amylases, and β-lactamase which were boiled in the presence of SDS and reducing agent prior to electrophoresis (*9, 12*). Intuitively speaking, the oligomeric enzymes whose functional unit is composed of two or more polypeptides differing by size are likely to be irreversibly inactivated by SDS and other denaturants, especially when the quaternary structure is prerequisite for activity. In this case, renaturation in gels after SDS-PAGE cannot occur because subunits will have moved to different positions in the gel. However, in some instances catalytically active subunit may be directly identified if it does not require other subunits for activity. For example, Paudel and Carlson (*16*) have identified the γ-subunit of phosphorylase kinase to be the catalytically active one because it showed activity on SDS gels by itself in the absence of α, β, and δ-subunits. Thus monomeric enzymes and oligomeric enzymes that are composed of monomers of the same molecular weight are more likely to be active or renature and suitable for study on SDS-gels by zymogram techniques. A variety of enzymes, especially those from extremely thermophilic bacteria, such as the malic enzyme of *Sulfolobus solfatoricus* (4), caldolysin of *Thermus aquaticus* (3), glyceraldehyde 3-phosphate dehydrogenase of *Thermus thermophilus* (2), α-glucosidase of *Bacillus thermoglusidius* (1), ATPase of thermophilic bacterium PS3 (6), and acid carboxypeptidase of *Penicillium janthinellum* (5), have been shown to be stable and /or active in the presence of varying concentrations (0.05% to 1%) of SDS. We predict that these enzymes and others from thermophilic microorganisms will be amenable to analysis on SDS gels by zymogram techniques. In fact, some of these enzymes have a temperature optimum around 75° C, and they may show activity on gels after samples are heated in the presence of SDS at high temperatures.

β-glucosidase zymograms on SDS gels and the stability and activity of these enzymes after SDS treatment or in the presence of SDS offer new approaches for the study of their structure and function. Applying the same approaches to other enzyme systems is not beyond the realm of possibility. Thus, the earlier studies (8-17, 19) and our data provide the basis for developing new approaches utilizing ionic detergents and other denaturants to study the quaternary structure of enzymes, investigate and define the roles of their individual subunit polypeptides, study the basis of other protein-to-protein interactions, and develop approaches and engineer enzymes for performing enzyme catalyzed reactions in denaturants or under denaturing conditions. β-glucosidase, being one of the key enzymes in the conversion of cellulose to glucose, is worthy of investigation for its potential in biomass conversion under denaturing conditions.

Future Research

Our data show that β-glucosidases are stable and active in the presence of protein denaturants to varying degrees depending on the source of the enzyme. However, the data do not permit any definitive conclusions either regarding the structural changes that β-glucosidases undergo in denaturants or reversibility of these structural changes. Future research should extend these studies to higher concentrations of the same as well as additional denaturants and employ physical techniques to investigate the nature of structural perturbations brought about during and after exposure to denaturants. In addition, pH, temperature, concentration, and time dependence of enzyme stability and activity in denaturants and their effects on kinetic parameters should be the subject of more thorough and systematic studies using pure enzymes from a variety of sources. The use of purified enzyme preparations, especially protease free enzyme preparations, is crucial to a meaningful interpretations of results.

Literature Cited

1. Suzuki, Y.; Yuki, T.; Kishigami, T.; Abe, S. *Biochim. Biophys. Acta.* **1976**, *445*, 386-397.
2. Fujita, S. C.; Oshima, T.; Imahori, K. *Eur. J. Biochem.* **1976**, *64*, 57-68.
3. Cowan, D. A.; Daniel, R. M. *Biochim. Biophys. Acta.* **1982**, *705*, 293-305.
4. Guagliardi, A.; Manco, G.; Rossi, M.; Bartolucci, S. *Eur. J. Biochem.* **1989**, *183*, 25-30.
5. Yokoyama, S.; Miyabe, T.; Oobayashi, A.; Tanabe, O.; Ichishima, E. *Agric. Biol. Chem.* **1977**, *41*, 1379-1383.
6. Norling, B. *Biochem. Biophys. Res. Comm.* **1986**, *136*, 899-905.
7. Ferdman, D. L.; Himmelreich, N. G.; Dyadyusha, G. P. *Biochim. Biophys. Acta.* **1970**, *219*, 372-378.
8. Rosenthal, A. L.; Lacks, S. A. *Anal. Biochem.* **1977**, *80*, 76-90.
9. Lacks, S. A.; Springhorn, S. S. *J. Biol. Chem..* **1980**, *255*, 7467-7473.
10. Spanos, A.; Sedgwick, S. G.; Yarranton, G. T.; Hubscher, U.; Banks, G. R. *Nucleic Acids Research* **1981**, *9*, 1825-1839.
11. Heussen, C.; Dowdle, E. G. *Anal. Biochem.* **1980**, *102*, 196-202.
12. Tai, P. C.; Zyk, N.; Citri, N. *Anal. Biochem.* **1985**, *144*, 199-203.
13. Geahlen, R. L.; Anostario, M.; Low, P. S.; Harrison, M. L. *Anal. Biochem.* **1986**, *153*, 151-158.
14. Nellaiappan, K.; Vinayagam, A. *Stain Tech.* **1986**, *61*, 269-272.
15. Schmidt, T. M.; Bleakley, B.; Nealson, K. H. *Appl. Env. Microb.* **1988**, *54*, 2793-2797.
16. Paudel, H. K.; Carlson, G. M. *Archiv. Biochem. Biophys.* **1988**, *264*, 641-646.
17. Audy, P.; Grenier J.; Asselin, A. *Comp. Biochem. Physiol.* **1989**, *92B*, 523-527.
18. Gillespie, J. P.; Bidochka, M. J.; Khachatourians, G. G. *Comp. Biochem. Physiol.* **1991**, *98C*, 351-358.
19. Esen, A.; Gungor, G. *Theoret. Appl. Electroph.* **1991**, *2*, 63-69.
20. Veronese, F. M.; Boccù, E.; Fontana, A. *Biochemistry.* **1976**, *15*, 4026-4033.
21. Sundaram, T. K.; Wright, I. P.; Wilkinson, A. R. *Biochemistry.* **1980.**, *19*, 2017-2022.
22. Veronese, F. M.; Boccù, E.; Schiavon, O.; Grandi, C.; Fontana, A. *J. Appl. Biochem.* **1984**, *6*, 39-47.
23. Laane, C.; Boeren, S.; Vos, K.; Veeger, C. *Biotechnol. Bioeng.* **1987**, *30*, 81-87.
24. Butler, L. G. *Enzyme Microb. Technol.* **1979**, *1*, 253-259.
25. Contaxis, C. C.; Reithel, F. J. *J. Biol. Chem.* **1971**, *246*, 677-685.
26. Zaks, A.; Klibanov, A. M. *J. Biol. Chem.* **1988**, *263*, 3194-3201.
27. Esen, A. *Plant Physiol.* **1992**, *98*, 174-182.
28. Tanford, C. *The Hydrophobic Effect: Formation of Micelles and Biological Membranes.* John Wiley & Sons, New York, N.Y.. **1980**.
29. Strambini, G. B.; Gonnelli, M. *Biochemistry* **1986**, *25*, 2471-2476.
30. Mashino, T.; Fridovich, I. *Arch. Biochem. Biophys.* **1987**, *258*, 356-360.

_____

31. Herskovits, T. T.; Gadegbeku, B.; Jaillet, H. *J. Biol. Chem.* **1970**, *245*, 2588-2598.
32. Esen, A.; Cokmus C. *Plant Sci.* **1991**, *74*, 17-26.
33. Stuber, C. W.; Goodman, M. M.; Johnson, F. M. *Biochem. Genet.* **1977**, *15*, 383-394.

RECEIVED March 24, 1993

Chapter 16

Use of Immobilized β-Glucosidase in the Hydrolysis of Cellulose

J. Woodward[1], L. J. Koran, Jr.[2,5], L. J. Hernandez[3,6], and L. M. Stephan[4,7]

[1]Chemical Technology Division, Oak Ridge National Laboratory, Oak Ridge, TN 37831–6194
[2]Department of Chemistry, University of Wisconsin, Stevens Point, WI 54481
[3]Department of Chemical Engineering, University of Puerto Rico, Mayaguez, PR 00680
[4]Kashunimiut School, Chevak, AK 99563

The supplementation of commercially produced *Trichoderma reesei* cellulases with ß-glucosidases, which possess high specific activity toward cellobiose, should prove useful for increasing the rate and extent of the hydrolysis of cellulosic substrates. Since cellobiose is soluble, ß-glucosidase could be used in an immobilized form and subsequently recovered and reused. We have prepared an immobilized ß-glucosidase (*Aspergillus niger*) by its entrapment within maintenance-free propylene glycol alginate/bone gelatin spheres. The enzyme thus immobilized is thermally stable at 40°C for several months, during which time it can be used for the continuous hydrolysis of cellobiose without loss of efficiency. The data indicate there is no loss of ß-glucosidase activity due to its leakage from the spheres. These biocatalytic spheres can also be dried (which makes them suitable for transporting) and subsequently rehydrated several times without any loss in catalytic activity. They have been used to supplement a commercial cellulase preparation for the hydrolysis of newsprint, recovered from the reaction mixture, and reused nine times without appreciable loss in activity or conversion of cellulose to glucose.

The enzyme ß-glucosidase (EC 3.2.1.21), also known by its trivial name cellobiase, catalyzes the hydrolysis of cellobiose, a dimer of glucose, to glucose (*1*). It is a vital component of the mixture of enzymes termed cellulase that catalyzes the hydrolysis of cellulose to glucose. Without sufficient ß-glucosidase present in cellulose, little glucose is formed; the main product then is cellobiose,

[5]Oak Ridge Science and Engineering Research Semester Student, 1991
[6]Oak Ridge Associated Universities Professional Intern, 1992
[7]U.S. Department of Energy Teacher Research Associate, 1992

which, in turn, inhibits the reaction so that cellulose hydrolysis is greatly impaired. It has been shown that supplementation of commercially produced cellulases from fungal sources such as *Trichoderma reesei* with ß-glucosidase produced by the fungus *Aspergillus niger* increases the rate and extent of glucose production (*2,3*). The native level of ß-glucosidase activity in cellulase is, therefore, insufficient for the maximum rate and extent of glucose production to be reached.

Since the substrate of ß-glucosidase, cellobiose, is water soluble, an immobilized (water-insoluble) ß-glucosidase preparation could be used to supplement commercial cellulase/cellulose mixtures, and such a preparation could be subsequently recovered and reused.

There have been many reports of immobilized ß-glucosidase preparations in the literature (e.g., see *4-7*), although no preparations are used commercially at the present time. In 1982, it was shown that *A. niger* ß-glucosidase, normally inhibited by cellobiose concentrations greater than 10 mM, was not, apparently, subject to inhibition by cellobiose concentrations as high as 100 mM if it were immobilized and entrapped within calcium alginate gel spheres (*8*). This was significant because of the possibility that concentrated cellulose-derived cellobiose solutions could be hydrolyzed efficiently (i.e., without inhibition) by such biocatalytic spheres. The disadvantage of using calcium alginate spheres is that they are not structurally stable in continuous-flow systems unless calcium ions are added.

The invention and development of maintenance-free propylene glycol alginate/bone gelatin (PGAG) spheres for the entrapment of microorganisms (*9*) eliminated this disadvantage, and they were used for the entrapment of ß-glucosidase (*10*), the first isolated biocatalyst to be entrapped within such spheres. These spheres were made structurally stable by soaking them in 0.1 N sodium hydroxide (NaOH) for at least 15 min; after such treatment, they required no further maintenance (i.e., addition of metal ions). However, storage in the NaOH solution, even for a few minutes, when used for entrapment, resulted in a dramatic decline in ß-glucosidase activity due to alkali inactivation (pH 12.0). Storage in NaOH for less time resulted in the formation of structurally unstable spheres.

There are two criteria that need to be met with regard to the development of gel sphere-entrapped ß-glucosidase prior to its use for the continuous and hence long-term use for the hydrolysis of cellulose-derived cellobiose solutions: (1) the gel spheres themselves must be structurally stable, and (2) the activity of the entrapped enzyme must be stable and not leak out of the spheres. The work reported here describes the chemical attachment of the ß-glucosidase to gelatin prior to the formation of the PGAG spheres, preventing leakage of the enzyme, and structural stabilization of the spheres by soaking them in a glutaraldehyde solution, rather than in NaOH, after their formation. The resulting spheres are structurally stable for several months, during which time the catalytic activity of ß-glucosidase does not decline to any great extent, and they can be used for the continuous hydrolysis of cellobiose to glucose during this time. The properties of ß-glucosidase thus immobilized are described.

Materials and Methods

ß-Glucosidase was a gift from Novo Nordisk Bioindustrials, Inc., Danbury, Connecticut. The product, Novozym 188, is produced by submerged fermentation of a selected strain of *Aspergillus niger*. Celluclast 1.5L (crude cellulase from *Trichoderma reesei*) was also a gift from Novo. Deionized bone gelatin was obtained from Rousselot, Paris, France; propylene glycol alginate (Kelcoloid S) was purchased from Kelco, Clark, New Jersey. Cellobiose and glutaraldehyde [25% (w/v) grade I] were purchased from Sigma Chemical Company, St. Louis, Missouri. Disposable PD-10 columns containing Sephadex G-25 M gel were purchased from Pharmacia LKB, Piscataway, New Jersey.

Preparation of Gel-Entrapped ß-Glucosidase. The enzyme was prepared as follows: 1 mL of crude enzyme was mixed with 4.0 mL of a solution of 50 mM sodium acetate at pH 5.0. Two aliquots, each consisting of 2.5 mL of this solution, were then filtered through a PD-10 column, and the filtered enzyme (total volume 7.0 mL) was used for gel entrapment. One gram of bone gelatin and 0.2567 g of cellobiose were dissolved in 44 mL of a solution of 50 mM sodium acetate at pH 5.0. The final cellobiose concentration was 15 mM. One mL of glutaraldehyde and 5.0 mL of the filtered enzyme were added to this solution, and the mixture was incubated at 23°C for 1 h. This was done to chemically cross-link ß-glucosidase to gelatin. The solution was then dialyzed at 4°C for 20 h to remove excess glutaraldehyde, and 25 mL of the dialyzed solution was then placed in a thermojacketed mixing apparatus. Two grams of propylene glycol alginate, 14.5 g of bone gelatin, and 75 mL of 50 mM sodium acetate, pH 5.0, were added, and the mixture was stirred at 40°C for 15 to 30 min until everything was dissolved.

The resulting viscous solution was then transferred to a gel sphere production apparatus and kept at 40°C. Spheres were formed by using 20 lb/in.² of gas pressure (helium or nitrogen) to force the viscous liquid through a needle as discrete droplets which fell into 500 mL of mineral oil on the surface of 200 mL of ice cold 50 mM sodium acetate, pH 5.0. One hundred mL of gel spheres, 2 mm in diameter, were formed. The spheres were removed from the oil/aqueous interface and stored for 20 h at 4°C in a 0.5% (w/v) solution of glutaraldehyde at pH 5.0. They were then washed 4X with 500 mL of distilled water and 3X with 200 mL of 50 mM sodium acetate, pH 5.0, and stored in the sodium acetate solution at pH 5.0.

Alternative Methodology. It should be noted that the amount of ß-glucosidase that could be chemically cross-linked to gelatin could, theoretically, be increased to as much as 50 mL of the filtered enzyme solution. Consequently, the amount of entrapped ß-glucosidase could be varied.

Hydrolysis of Newsprint. Local newsprint (~70% cellulose), 0.1 g, was incubated at 40°C in a 10-mL volume of 50 mM sodium acetate buffer, pH 5.0, containing 100 μl of gel-filtered (Sephadex G-25) celluclast (diluted 5X) and 1.0 mL of ß-glucosidase PGAG spheres. At intervals of time, glucose production was monitored using the hexokinase assay reagent. It should be noted that the spheres were encased by an elastic nylon mesh bag to physically isolate them from the newsprint, making their recovery easy.

Results and Discussion

Properties and Characteristics of Gel-Entrapped ß-Glucosidase.

Catalytic Activity. The activity of the filtered ß-glucosidase was determined to be approximately 170 units/mL, where 1 unit is defined as that amount of enzyme required to produce 1 μmol glucose/min from 10 mM cellobiose at 40°C in 50 mM sodium acetate, pH 5.0. Monitoring the activity during the procedures involved in the entrapment of ß-glucosidase within propylene glycol alginate/bone spheres was done, and the results are shown in Table I (for more details about the assay, see reference 10). The data indicate that after cross-linking ß-glucosidase to gelatin, followed by dialysis, there was only a 10% loss in activity. There was a negligible loss in activity from the spheres into the sodium acetate solution (supernatant) after sphere formation. The total activity of the spheres represented only 27% of the original activity. This apparent loss in activity may be caused by substrate diffusion and/or by the soaking of the spheres in glutaraldehyde overnight.

Table I. Entrapment of ß-Glucosidase in PGAG Spheres*

Initial Activity	Activity After Cross-linking	Activity in Supernatant	Entrapped Activity
877	790	0	238

*Values refer to total activity (i.e., 100 mL of gel spheres contained 238 units of cellobiase activity).

pH and Temperature Activity Profiles. The pH activity profile of the entrapped enzyme was shifted slightly more toward the alkaline side (Figure 1a), the pH optima of the native and entrapped enzyme being 4.5 and 5.0, respectively. The reason for this is unknown but is probably related to an unequal distribution of hydrogen and hydroxyl ions between the gelatin (to which ß-glucosidase is attached), making up the bulk of the sphere, and the external solution (*11*).

The temperature activity profiles of the native and entrapped enzyme indicated that there was no shift in temperature optimum (Figure 1b). The energy of activation (E_a) was calculated for the native and immobilized enzymes at the different concentrations of cellobiose used in the assay (Figure 2). A detailed mathematical explanation for the findings, that E_a of this ß-glucosidase is not only a function of substrate concentration but is also lower in the case of the entrapped enzyme, is beyond the scope of this work. It is very likely that in

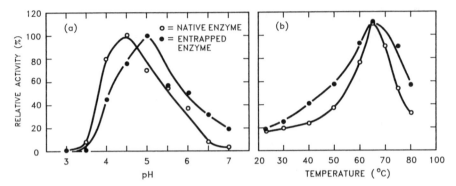

Figure 1. The effect of pH (a) and temperature (b) on the activity of native and gel-entrapped ß-glucosidase. Sodium acetate (50 mM) and sodium phosphate (50 mM) buffers were used to obtain pH 3.0 to 6.0 and pH 7.0, respectively.

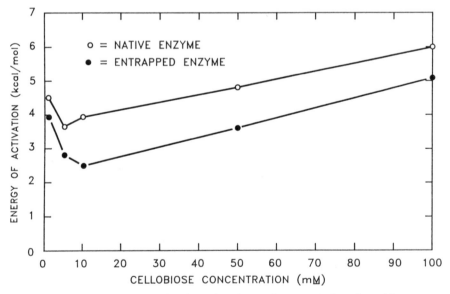

Figure 2. The energy of activation of native and entrapped ß-glucosidase as a function of substrate concentration.

the case of the immobilized enzyme, intraparticle mass-transfer resistance plays a key role (*12*).

Effect of Substrate Concentration. It has been previously reported that calcium alginate gel-entrapped ß-glucosidase is not subject to substrate inhibition, unlike the native enzyme (*8,10*). This was also found to be the case for this enzyme entrapped within the PGAG spheres (Figure 3). The cellobiase activity of the spheres (~2.5 units/mL spheres) was virtually unchanged between a substrate (cellobiose) concentration of 10-190 mM [0.3422-6.5% (w/v)].

Effect of Drying/Rehydration on Activity. The hydrated spheres, as shown in Figure 4A, were ~2 mm in diameter and could be dried in air at 23°C, resulting in hard pellets that were 1 mm in diameter (Figure 4B). It would be practically advantageous, if these biocatalytic spheres became commercially available, to ship them in dry form. Once rehydrated, it would be essential that they retained their activity. Firstly, it is important to note that these spheres can be dehydrated and rehydrated repeatedly. This is in contrast to calcium alginate and κ-carrageenan gel spheres which, once dehydrated, cannot be rehydrated. Secondly, the spheres were dehydrated/rehydrated 5 times with approximately 20% loss in initial sphere activity. Rehydration of the spheres was accomplished by soaking them in 50 mM sodium acetate, pH 5.0, for 30 min.

The practical implications of these data and observations are that the spheres could be manufactured, dried, packaged, and shipped out to a customer who could then rehydrate, use, and dry them again until required again. The shelf life of these spheres and the ß-glucosidase activity within them can probably be measured in years. The dehydrated spheres are not hygroscopic when sitting out on the bench top within a climate-controlled laboratory. We did not measure the pore size of the spheres.

Thermal Stability of the Biocatalytic Spheres at 40°C. The biocatalytic spheres containing ß-glucosidase activity were stirred at 40°C in 50 mM sodium acetate, pH 5.0, and 0.1% sodium azide as a preservative/bacterial growth inhibitor. There was little loss in activity after 74 d (Figure 5). These data are significant for two reasons: (1) the catalytic activity of the spheres is very stable under these conditions, and (2) there cannot be any significant leakage of ß-glucosidase from the spheres during this time. Otherwise, the activity of the spheres would have been expected to decline. Also, no significant activity was measured in the solution in which the spheres were stirred, indicating that leakage of enzyme did not occur. It should also be noted that the entrapped enzyme exhibited greater thermal stability than the native enzyme at higher temperatures. For example, when native and entrapped enzymes were heated in the presence of cellobiose (1 mM) at 80°C, after 30 min, there were 9 and 25% recoveries of their maximum rate of hydrolysis, respectively.

The higher thermal stability of the entrapped enzyme may be due, in part, to the thermal stabilization of ß-glucosidase by its incubation with glutaraldehyde that we and others have previously observed (*13-15*). Thermal stabilization of native and entrapped ß-glucosidase by cellobiose would be likely. Little loss of

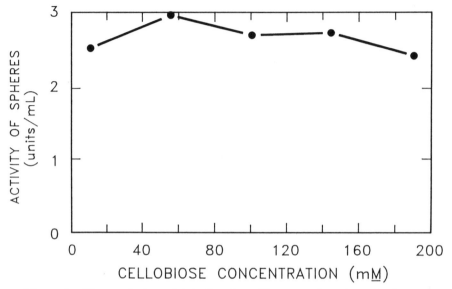

Figure 3. The catalytic activity of ß-glucosidase as a function of substrate concentration.

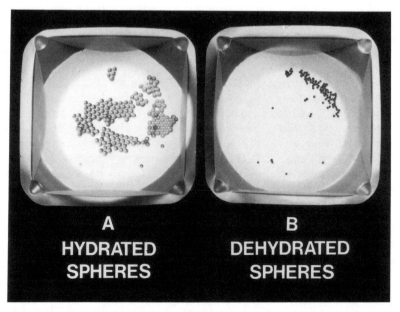

Figure 4. Appearance of hydrated and dehydrated PGAG gel spheres containing ß-glucosidase.

catalytic activity was noted when the entrapped enzyme was stored at 4°C for 7 months.

Continuous Hydrolysis of Cellobiose.   Thirty-three mL of the biocatalytic spheres containing ß-glucosidase (~80 units of activity) was packed in a thermojacketed, tapered bioreactor at 40°C, total volume 50 mL, and a solution of cellobiose [10 mM or 0.342% (w/v) in 50 mM sodium acetate, pH 5.0, containing 0.1% sodium azide] was pumped upward through the column at a flow rate of 0.2 mL/min, and the glucose concentration of the effluent was determined.   There was a 100% conversion of a 10 mM cellobiose solution to glucose which was maintained over a period of 24 d.   The cellobiose concentration was increased to 100 mM after this time.   The percentage conversion was 80-90% over a period of 74 d (Figure 6).   These data also provide evidence that the biocatalytic spheres containing ß-glucosidase are catalytically stable and do not lose their catalytic activity through leakage over time.

Miscellaneous Observations.   The 2% (w/v) propylene alginate glycol used in the manufacture of the spheres is essential to maintaining their structural integrity prior to their being soaked in glutaraldehyde.   Propylene glycol alginate and bone gelatin are not inhibitors of cellobiase activity.

Use of Entrapped ß-Glucosidase in the Hydrolysis of Newsprint.   Local newsprint was hydrolyzed using a commercial cellulase preparation (celluclast 1.5L) in the presence and absence of entrapped ß-glucosidase.   The data (Figure 7) indicate the expected enhancement of the rate and extent of the hydrolysis of the cellulose component of newsprint to glucose when the entrapped enzyme was added to the reaction mixture.   The entrapped enzyme was recovered from the reaction mixture, washed, soaked in buffer overnight to remove any remaining glucose from the interior of the sphere, and reused.   The newsprint hydrolysis experiment was repeated nine times with, basically, the same result.

Conclusions

The methodology we have described for the entrapment of a commercially available ß-glucosidase preparation may prove useful for the entrapment of enzymes in general.   The spheres are structurally stable (without the addition of exogenously added metal ions) and can be used in a continuous-flow system for months.   The activity of ß-glucosidase entrapped within PGAG spheres is also stable and can be used repeatedly for the hydrolysis of cellulose-derived cellobiose to glucose.   Other potential uses for the entrapped enzyme included the synthesis (esterification) of glucosides of physiological importance (in organic media?), as well as the synthesis of cello-oligosaccharides.

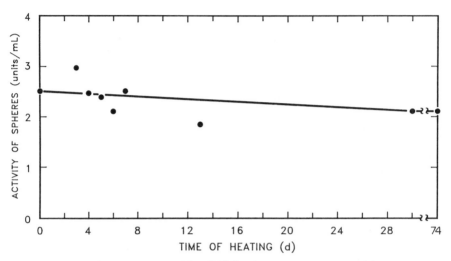

Figure 5. Thermal stability (40°C) of entrapped ß-glucosidase.

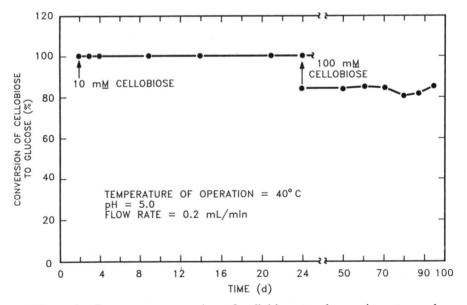

Figure 6. Percentage conversion of cellobiose to glucose in a tapered continuous-flow bioreactor by entrapped ß-glucosidase.

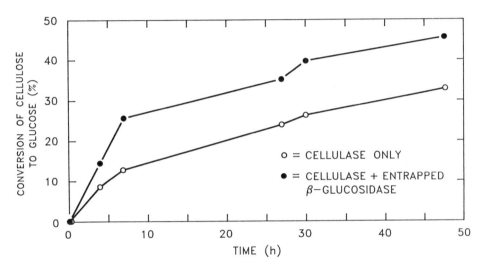

Figure 7. Percentage conversion of cellulose (in newsprint) to glucose by cellulase in the absence and presence of entrapped ß-glucosidase.

Acknowledgments

The authors thank J. M. Cosgrove and M. E. Reeves for reviewing the manuscript and D. J. Weaver and S. A. Hoglund for secretarial assistance. This work was supported by the Chemical Sciences Division, Office of Basic Energy Sciences, U.S. Department of Energy under contract DE-AC05-84OR21400 with Martin Marietta Energy Systems, Inc.

Literature Cited

1. Woodward, J.; Wiseman, A. *Enzyme Microb. Technol.* **1982**, *4*, 73-78.
2. Sternberg, D.; Vijayakumar, P.; Reese, E. T. *Can. J. Microbiol.* **1977**, *23*, 139-147.
3. Woodward, J.; Bales, J. C. In *Bioproducts and Bioprocesses*; Fiechter, A., Okada, H., Tanner, R. D. Eds.; Springer: Berlin, 1989, pp 87-101.
4. Sundstrom, D. W.; Klei, H. E.; Coughlin, R. W.; Biederman, G. J., Grouwer, C. A. *Biotechnol. Bioeng.* **1981**, *23*, 473-485.
5. Woodward, J.; Wohlpart, D. L. *J. Chem. Tech. Biotechnol.* **1982**, *32*, 547-552.
6. Bisset, F.; Sternberg, D. *Appl. Environ. Microbiol.* **1978**, *35*, 750-755.
7. Woodward, J. *J. Biotechnol.* **1989**, *11*, 299-312.
8. Lee, J. M.; Woodward, J. *Biotechnol. Bioeng.* **1983**, *25*, 2441-2451.
9. Scott, C. D.; Woodward, C. A.; Byers, C. H. (1991) U.S. Patent 4,995,985.
10. Woodward, J.; Clarke, K. M. *Appl. Biochem. Biotechnol.* **1991**, *28/29*, 277-283.
11. Goldstein, L.; Levin, Y.; Katchalski, E. *Biochemistry* **1964**, *3*, 1913-1919.
12. Lee, J. M. *Biochemical Engineering*; Prentice-Hall: Englewood Cliffs, NJ, **1992**.
13. Woodward, J.; Whaley, K. S.; Zachry, G. S.; Wohlpart, D. L. *Biotechnol. Bioeng. Symp.* **1981**, *11*, 619-629.
14. Baker, J. O.; Oh, K. K.; Grohmann, K.; Himmel, M. E. *Biotechnol. Lett.* **1988**, *10*, 325-330.
15. Woodward, J.; Kapps, K. M. *Appl. Biochem. Biotechnol.* **1992**, *34/35*, 341-347.

RECEIVED February 1, 1993

INDEXES

Author Index

Affiliation Index

Subject Index

Production: Margaret J. Brown
Indexing: Deborah H. Steiner
Acquisition: Rhonda Bitterli
Cover design: Amy Hayes

Printed and bound by Maple Press, York, PA

Bestsellers from ACS Books

The ACS Style Guide: A Manual for Authors and Editors
Edited by Janet S. Dodd
264 pp; clothbound ISBN 0–8412–0917–0; paperback ISBN 0–8412–0943–X

The Basics of Technical Communicating
By B. Edward Cain
ACS Professional Reference Book; 198 pp;
clothbound ISBN 0–8412–1451–4; paperback ISBN 0–8412–1452–2

Chemical Activities (student and teacher editions)
By Christie L. Borgford and Lee R. Summerlin
330 pp; spiralbound ISBN 0–8412–1417–4; teacher ed. ISBN 0–8412–1416–6

Chemical Demonstrations: A Sourcebook for Teachers,
Volumes 1 and 2, Second Edition
Volume 1 by Lee R. Summerlin and James L. Ealy, Jr.;
Vol. 1, 198 pp; spiralbound ISBN 0–8412–1481–6;
Volume 2 by Lee R. Summerlin, Christie L. Borgford, and Julie B. Ealy
Vol. 2, 234 pp; spiralbound ISBN 0–8412–1535–9

Chemistry and Crime: From Sherlock Holmes to Today's Courtroom
Edited by Samuel M. Gerber
135 pp; clothbound ISBN 0–8412–0784–4; paperback ISBN 0–8412–0785–2

Writing the Laboratory Notebook
By Howard M. Kanare
145 pp; clothbound ISBN 0–8412–0906–5; paperback ISBN 0–8412–0933–2

Developing a Chemical Hygiene Plan
By Jay A. Young, Warren K. Kingsley, and George H. Wahl, Jr.
paperback ISBN 0–8412–1876–5

Introduction to Microwave Sample Preparation: Theory and Practice
Edited by H. M. Kingston and Lois B. Jassie
263 pp; clothbound ISBN 0–8412–1450–6

Principles of Environmental Sampling
Edited by Lawrence H. Keith
ACS Professional Reference Book; 458 pp;
clothbound ISBN 0–8412–1173–6; paperback ISBN 0–8412–1437–9

Biotechnology and Materials Science: Chemistry for the Future
Edited by Mary L. Good (Jacqueline K. Barton, Associate Editor)
135 pp; clothbound ISBN 0–8412–1472–7; paperback ISBN 0–8412–1473–5

For further information and a free catalog of ACS books, contact:
American Chemical Society
Distribution Office, Department 225
1155 16th Street, NW, Washington, DC 20036
Telephone 800–227–5558

Highlights from ACS Books

Good Laboratory Practice Standards: Applications for Field and Laboratory Studies
Edited by Willa Y. Garner, Maureen S. Barge, and James P. Ussary
ACS Professional Reference Book; 572 pp; clothbound ISBN 0–8412–2192–8

Silent Spring Revisited
Edited by Gino J. Marco, Robert M. Hollingworth, and William Durham
214 pp; clothbound ISBN 0–8412–0980–4; paperback ISBN 0–8412–0981–2

The Microkinetics of Heterogeneous Catalysis
By James A. Dumesic, Dale F. Rudd, Luis M. Aparicio, James E. Rekoske,
and Andrés A. Treviño
ACS Professional Reference Book; 316 pp; clothbound ISBN 0–8412–2214–2

Helping Your Child Learn Science
By Nancy Paulu with Margery Martin; Illustrated by Margaret Scott
58 pp; paperback ISBN 0–8412–2626–1

Handbook of Chemical Property Estimation Methods
By Warren J. Lyman, William F. Reehl, and David H. Rosenblatt
960 pp; clothbound ISBN 0–8412–1761–0

Understanding Chemical Patents: A Guide for the Inventor
By John T. Maynard and Howard M. Peters
184 pp; clothbound ISBN 0–8412–1997–4; paperback ISBN 0–8412–1998–2

Spectroscopy of Polymers
By Jack L. Koenig
ACS Professional Reference Book; 328 pp;
clothbound ISBN 0–8412–1904–4; paperback ISBN 0–8412–1924–9

Harnessing Biotechnology for the 21st Century
Edited by Michael R. Ladisch and Arindam Bose
Conference Proceedings Series; 612 pp;
clothbound ISBN 0–8412–2477–3

From Caveman to Chemist: Circumstances and Achievements
By Hugh W. Salzberg
300 pp; clothbound ISBN 0–8412–1786–6; paperback ISBN 0–8412–1787–4

The Green Flame: Surviving Government Secrecy
By Andrew Dequasie
300 pp; clothbound ISBN 0–8412–1857–9

For further information and a free catalog of ACS books, contact:
American Chemical Society
Distribution Office, Department 225
1155 16th Street, NW, Washington, DC 20036
Telephone 800–227–5558